Collection of problems
in probability theory

L. D. MESHALKIN
Moscow State University

Collection of problems in probability theory

Translated from the Russian and edited by

LEO F. BORON
University of Idaho

and

BRYAN A. HAWORTH
University of Idaho and
California State College, Bakersfield

NOORDHOFF INTERNATIONAL PUBLISHING, LEYDEN

© 1973 Noordhoff International Publishing, Leyden, The Netherlands
Softcover reprint of the hardcover 1st edition 1973

ISBN-13:978-94-010-2360-3   e-ISBN-13:978-94-010-2358-0
DOI: 10.1007/978-94-010-2358-0

Library of Congress Catalog Card Number: 72–76789

Original title "Sbornik zadach po teorii veroyatnostey" published in 1963 in Moscow

Contents

Contents

4   Basic limit theorems   56

5   Characteristic and generating functions   71

6   Application of measure theory   82

Editor's foreword

The Russian version of *A collection of problems in probability theory* contains a chapter devoted to statistics. That chapter has been omitted in this translation because, in the opinion of the editor, its content deviates somewhat from that which is suggested by the title: problems in probability theory.

The original Russian version contains some errors; an attempt was made to correct all errors found, but perhaps a few still remain.

An index has been added for the convenience of the reader who may be searching for a definition, a classical problem, or whatever. The index lists pages as well as problems where the indexed words appear.

The book has been translated and edited with the hope of leaving as much "Russian flavor" in the text and problems as possible. Any peculiarities present are most likely a result of this intention.

August, 1972   Bryan A. Haworth

Foreword to the
Russian edition

This *Collection of problems in probability theory* is primarily intended for university students in physics and mathematics departments. Its goal is to help the student of probability theory to master the theory more profoundly and to acquaint him with the application of probability theory methods to the solution of practical problems. This collection is geared basically to the third edition of the GNEDENKO textbook *Course in probability theory*, Fizmatgiz, Moscow (1961), *Probability theory*, Chelsea (1965). It contains 500 problems, some suggested by monograph and journal article material, and some adapted from existing problem books and textbooks. The problems are combined in nine chapters which are equipped with short introductions and subdivided in turn into individual sections. The problems of Chapters 1–4 and part of 5, 8 and 9 correspond to the semester course *Probability theory* given in the mechanics and mathematics department of MSU. The problems of Chapters 5–8 correspond to the semester course *Supplementary topics in probability theory*. Difficult problems are marked with an asterisk and are provided with hints. Several tables are adjoined to the collection. Answers are given only to odd numbered problems.

This is done to train the student to evaluate independently the correctness of a solution, and also so that the material of the collection could be used for supervised work.

To supplement the collection, the teacher can make use of the following three problem books which contain well chosen material on statistics and the theory of stochastic processes:

1. VOLODIN, B. G., M. P. GANIN, I. YA. DINER, L. B. KOMAROV, A. A. SVESHNIKOV, and K. B. STAROBIN. *Textbook on problem solving*

in probability theory for engineers. Sudpromgiz, Leningrad (1962).

2. LAJOS TAKÁCS. *Stochastic processes. Problems and solutions.* Wiley (1970) (in the series Methuen's monographs on applied probability and statistics).

3. DAVID, F. N. and E. S. PEARSON. *Elementary statistical exercises.* Cambridge University Press (1961).

My co-workers and degree candidates of the MSU Department of Probability Theory were of enormous help in choosing and formulating these exercises. I am deeply indebted to them for this. In particular I wish to thank M. Arato, B. V. Gnedenko, R. L. Dobrushin and Ya. G. Sinai.

July 9, 1963   L. D. Meshalkin

1   *Fundamental concepts*

The problems of this chapter correspond basically to the material of sections 1–8 of B. V. GNEDENKO's textbook *The theory of probability,* Chelsea Publishing Co. (1967). We illustrate here for the sake of convenience the interrelations among events used in the sequel.

Suppose that a point in the plane is selected at random and that the events A and B consist of this point lying in the circle A or in the circle B respectively. In Figures 1, a)–1, e) the regions are shaded such that a point falling into a shaded region corresponds respectively to the events:

$$A \cup B, \quad A \cap B, \quad A \bigtriangleup B, \quad A - B, \quad \bar{A}.$$

In the usual set-theoretic terminology, these events are respectively called: in case a), the *union* of the events A and B; in case b), the *intersection* of the events A and B; in case c) the *symmetric difference* of events A and B; in case d), the *difference* of the events A and B; in case e), the *negation* or *complement* of the event A. We note that the event $A \bigtriangleup B$ is realized if and only if one and only one of the events A and B is realized. Figure 1, f) corresponds to the relation $B \subseteq A$. Figure 1, g) corresponds to the relation $A \cap B = \emptyset$, where $\emptyset$ denotes the empty set. If $A \cap B = \emptyset$, then A and B are said to be *incompatible* or *nonintersecting events*.

The problems of the second section are intended for those who are primarily interested in applying the theory to statistics. In these problems we use the following notation: N is the total number of objects under consideration; $N\{\ \}$ is the number of these objects having the property appearing in the braces. These problems have been adapted from the first chapter of the book *An introduction to the theory of statistics* by G. U. YULE and M. G. KENDALL, C. Griffin and Co., London (1937).

Starting with problem 23 it is assumed that the reader is familiar with the following aspects of probability:

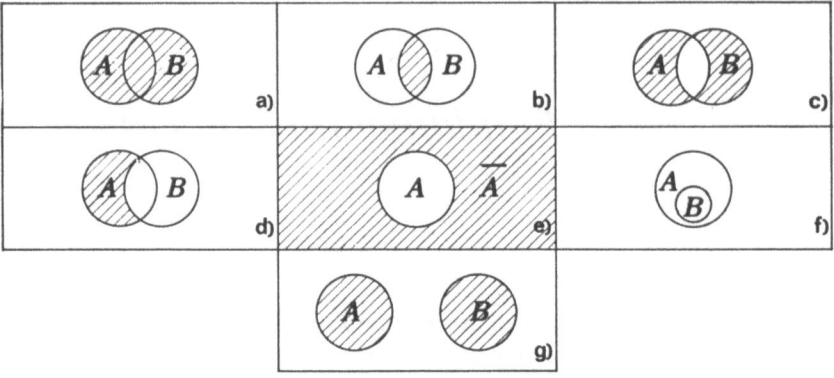

Fig. 1.

Let $\mathscr{E}$ be an experiment and let $\mathscr{A}$ be the collection of all possible outcomes of $\mathscr{E}$. Let S be the class of all subsets of $\mathscr{A}$. S is called the *sample space* associated with $\mathscr{E}$. An *event* is any subset of S. A *probability* is a set function P defined on S having the following properties:

(1) $P\{A\} \geqslant 0$ for all $A \in S$.

(2) If E is the set containing all possible outcomes then $P\{E\} = 1$.

(3) If $A = \bigcup_{i=1}^{\infty} A_i$, where $A_i \cap A_j = \emptyset$ $(i \neq j)$, then $P\{A\} = \sum_{i=1}^{\infty} P\{A_i\}$.

In many combinatorial problems it is very convenient to use the *classical definition of probability*. Suppose that as the result of a trial only one of n pairwise incompatible and equally probable results E_i $(i = 1, 2, ..., n)$ can be realized. We shall assume that the event A consists of m elementary results E_k. Then, according to the classical definition of probability,

$$P\{A\} = \frac{m}{n}.$$

The basic difficulty in solving problems by this method consists in a suitable choice of the space of elementary events. In this connection, particular attention must be given to verifying that the chosen elementary events are equally probable and that in the computation of m and n the same space of elementary events is used.

The simplest problems on arrangements acquaint one with certain applications of combinatorial methods in statistical physics. The term "statistics" just introduced is used in the sense specified for physics.

Almost all the problems of this section have been adapted from W. FELLER, *An introduction to probability theory and its applications*, Vol. I, Third Edition, Copyright © 1968 by John Wiley & Sons, Inc. One should recall the following facts. If, from among n objects, r are chosen, then the total number of possible combinations which might be obtained is $\binom{n}{r} = C_n^r$. The total number of permutations of n objects is $n!$ and the number of permutations (ordered samples without replacement) of size r from n objects is

$$n(n-1)\ldots(n-r+1) = P_n^r = r!C_n^r.$$

Special attention must be given to the problems of 1.6, geometric probability, for the solution of which sketches are particularly helpful. It is natural to introduce in these problems the concepts of distribution functions and density functions. More difficult problems from geometric probability can be found in 2.3, continuous distributions.

Geometric probability is defined in the following way: if some region R (in E^2 for example) is given, the probability that a point randomly located in R falls in some subregion R_0 of R is given by the ratio

$$\frac{\text{measure of } R_0}{\text{measure of } R} \left(\text{in } E^2, \frac{\text{area of } R_0}{\text{area of } R} \right).$$

Problems 60–70 go somewhat outside the framework of the obligatory course (in Soviet universities). They indicate the interrelationships of the above-introduced concepts with the problem of the metrization of a space with measure and linearly ordered sets. The material for these problems was adapted from the article by FRANK RESTL, published in the journal *Psychometrics 24*, No. 3 (1959) pp. 207–220. One should consult [2], [3] and [11] for additional reading.

1.1   Field of events

1. From among the students gathered for a lecture on probability theory one is chosen at random. Let the event A consist in that the chosen student is a young man, the event B in that he does not smoke, and the event C

in that he lives in the dormitory.

    a) Describe the event $A \cap B \cap C$.

    b) Under what conditions will the identity $A \cap B \cap C = A$ hold?

    c) When will the relation $\bar{C} \subseteq B$ be valid?

    d) When does the equation $\bar{A} = B$ hold? Will it necessarily hold if all the young men smoke?

2. A target consists of five discs bounded by concentric circles with radii r_k $(k = 1, 2, ..., 10)$, where $r_1 < r_2 < \cdots < r_{10}$. The event A consists in falling into the disc of radius r. What do the following events signify?

$$B = \bigcup_{k=1}^{6} A_k; \quad C = \bigcap_{k=1}^{10} A_k; \quad D = A_5 \triangle A_6; \quad E = \bar{A}_1 \cap A_2.$$

3. Prove that for arbitrary events A and B the relations $A \subset B$, $\bar{A} \supset \bar{B}$; $A \cup B = B$, $A \cap \bar{B} = \emptyset$ are equivalent.

4. Prove the following equalities:

    a) $\overline{A \cap B} = A \cup B$

    b) $\overline{\bar{A} \cup \bar{B}} = A \cap B$

    c) $A \cup B = (A \cap B) \cup (A \triangle B)$

    d) $\overline{A \triangle B} = (A \cap B) \cup (\bar{A} \cap \bar{B})$

    e) $A \triangle B = (\overline{A \cap \bar{B}}) \triangle (\overline{\bar{A} \cap B})$

    f) $\overline{\bigcup_{i=1}^{n} A_i} = \bigcap_{i=1}^{n} \bar{A}_i$

    g) $\overline{\bigcap_{i=1}^{n} A_i} = \bigcup_{i=1}^{n} \bar{A}_i.$

5. Prove that $A \triangle B = C \triangle D$ implies that $A \triangle C = B \triangle D$.

6. Prove that $\overline{(A \cup B)} \cap C = (\bar{A} \cap C) \cup (\bar{B} \cap C)$ holds if and only if $A \cap C = B \cap C$.

7. Prove that $A \triangle B \subseteq C$ implies that $A \subseteq (B \triangle C)$ if and only if $A \cap B \cap C = \emptyset$.

8. A worker made n parts. Let the event A_i $(i = 1, 2, ..., n)$ consist in that the i-th part made is defective. List the events consisting of the following:

    a) none of the parts is defective;

    b) at least one of the parts is defective;

    c) only one of the parts is defective;

    d) not more than two of the parts are defective;

    e) at least two parts are not defective;

    f) exactly two parts are defective.

9. Let A_n be the event that, at the n-th iteration of the experiment ε, the outcome A is realized; let $B_{n,m}$ be the event that in the first n repetitions of ε the outcome A is realized m times.

    a) Express $B_{4,2}$ in terms of the A_i.

    b) Interpret the event $B_m = \bigcup_n \{\bigcap_{k \geqslant n} B_{k,m}\}$.

    c) Are the relations $\bigcap_{n=1}^{\infty} A_n \subseteq \bar{B}$ and $\bigcap_{n=1}^{\infty} \bar{A}_n \subseteq B$, where $B = \bigcup_{m=1}^{\infty} B_m$ valid?

10. From the set E of points ω there are selected n subsets $A_i (i=1, 2, ..., n)$. For an arbitrary ω-set we define $\chi_C(\omega)$, the characteristic function of the set C, by setting $\chi_C(\omega) = 1$ if $\omega \in C$ and $\chi_C(\omega) = 0$ otherwise. Prove that, using A_i, one can construct sets B_k $(k = 1, 2, ..., 2^n)$ such that for an arbitrary bounded function

$$F(\omega) = F(\chi_{A_1}(\omega), ..., \chi_{A_n}(\omega))$$

there exist constants C_k such that

$$F(\omega) = \sum_k C_k \chi_{B_k}(\omega).$$

1.2   Interrelationships among cardinalities of sets

11. Prove that

    a) $N\{A \cap B\} + N\{A \cap C\} + N\{B \cap C\} \geqslant N\{A\} + N\{B\} + N\{C\} - N;$

    b) $N\{A \cap B\} + N\{A \cap C\} - N\{B \cap C\} \leqslant N\{A\}.$

12. In what sense may the inequality $N\{A \cap B\}/N\{B\} > N\{A \cap \bar{B}\}/N\{\bar{B}\}$ be interpreted as stating that Property B "favors" Property A. Show that, if B favors A, then A favors B.

13. If $N\{A\} = N\{B\} = \frac{1}{2}N$, prove that $N\{A \cap B\} = N\{\bar{A} \cap \bar{B}\}$.

14. If $N\{A\} = N\{B\} = N\{C\} = \frac{1}{2}N$ and $N(A \cap B \cap C) = N(\bar{A} \cap \bar{B} \cap \bar{C})$, show that $2N\{A \cap B \cap C\} = N\{A \cap B\} + N\{A \cap C\} + N\{B \cap C\} - \frac{1}{2}N.$

15. Show that the following data are incompatible:

$$N=1000; \quad N\{A \cap B\}=42$$
$$N\{A\}=525; \quad N\{A \cap C\}=147$$
$$N\{B\}=312; \quad N\{B \cap C\}=86$$
$$N\{C\}=470; \quad N\{A \cap B \cap C\}=25$$

Hint: Calculate $N\{\bar{A} \cap \bar{B} \cap \bar{C}\}$.

16. In a certain calculation the following numbers were given as those actually observed: $N=1000$; $N\{A\}=510$; $N\{B\}=490$; $N\{C\}=427$; $N\{A \cap B\}=189$; $N\{A \cap C\}=140$; $N\{B \cap C\}=85$. Show that they must contain some error or misprint and that possibly the misprint consists in omitting 1 before 85, given as the value of $N\{B \cap C\}$.

17. *Puzzle problem* (Lewis Carroll, *A Tangled Tale*, 1881). In a fierce battle, not less than 70% of the soldiers lost one eye, not less than 75% lost one ear, not less than 80% lost one hand and not less than 85% lost one leg. What is the minimal possible number of those who simultaneously lost one eye, one ear, one hand and one leg?

18. Show that if $N\{A\}=Nx$; $N\{B\}=2Nx$; $N\{C\}=3Nx$, $N\{A \cap B\}= =\{A \cap C\}=N\{B \cap C\}=Ny$, then the values of x and y cannot exceed $\frac{1}{4}$.

19. The investigator of a market reports the following data. Of 1000 persons questioned, 811 liked chocolates, 752 liked bonbons and 418 liked lollipops, 570 chocolates and bonbons, 356 chocolates and lollipops, 348 bonbons and lollipops, and 297 all three types of sweets. Show that this information contains an error.

20. The following data are the number of boys with certain groups of deficiencies per 10,000 boys of school age observed: A – deficiency in physical development, B – signs of nervousness, D – mental weakness.

$$N=10,000; \quad N\{D\}=789;$$
$$N\{A\}=877; \quad N\{A \cap B\}=338;$$
$$N\{B\}=1086; \quad N\{B \cap D\}=455.$$

Show that there are certain mentally retarded boys who display no deficiencies in physical development; determine the minimal number of these consistent with the data.

21. The following numbers are the analogous data for girls (see the preceding problem):

$$N=10,000; \quad N\{D\}=689;$$

$$N\{A\}=682; \qquad N\{A \cap B\}=248;$$
$$N\{B\}=850; \qquad N\{B \cap D\}=368.$$

Show that some physically undeveloped girls are not mentally retarded and determine the minimal number of them.

22. A coin was triply tossed 100 times; after each toss, the result was noted – either heads or tails. In 69 of the 100 cases, heads came up in the first toss; in 49 cases, heads in the second toss; in 53 cases, heads in the third toss. In 33 cases, heads came up in the first and second tosses and in 21 cases in the second and third. Show that there can be at least 5 cases in which heads occurred in all three tosses and that there cannot be more than 15 cases when for all three tosses a tail would occur, although not even one such case must necessarily occur.

1.3 Definition of probability

23. Given $p=P(A)$, $q=P(B)$, $r=P(A \cup B)$, find $P(A \bigtriangleup B)$, $P(A \cap \bar{B})$, $P(\bar{A} \cap \bar{B})$.

24. It is known that $P(A \cap B)=P(A)P(B)$(i.e., the events A and B are independent), $C \supset A \cap B$ and $\bar{C} \supset (\bar{A} \cap \bar{B})$. Prove that $P(A \cap C) \geqslant \geqslant P(A)P(C)$.

25. a) It is known that the simultaneous occurrence of the events A_1 and A_2 necessarily forces the occurrence of the event A; prove that

$$P(A) \geqslant P(A_1) + P(A_2) - 1.$$

b) Prove the following inequality for three events: if $A_1 \cap A_2 \cap A_3 \subset A$, then

$$P(A) \geqslant P(A_1) + P(A_2) + P(A_3) - 2.$$

26. In an experiment ε, three pairwise incompatible outcomes A_n are possible; also in the experiment ε, four other pairwise incompatible outcomes B_m are possible. The following compatible probabilities are known:

$$p_{11}=0.01, \qquad p_{21}=0.02, \qquad p_{31}=0.07,$$
$$p_{12}=0.02, \qquad p_{22}=0.04, \qquad p_{32}=0.15,$$

7

$$p_{13}=0.03, \qquad p_{23}=0.08, \qquad p_{33}=0.20,$$
$$p_{14}=0.04, \qquad p_{24}=0.06, \qquad p_{34}=0.28.$$

Find $P(A_n)$ and $P(B_m)$ for all n and m. (Also see exercise 83.)

27. A coin is tossed until it comes up with the same side twice in succession. To each possible outcome requiring n tosses, we ascribe the probability 2^{-n}. Describe the space of elementary events. Find the probability of the following events:

    a) the experiment ends at the sixth toss;

    b) an even number of tosses are required.

28. Two dice are thrown. Let A be the event that the total number of eyes is odd; B the event that at least one of the dice comes up a unit. Describe the events $A \cap B$, $A \cup B$, $\overline{A \cap B}$. Find their probabilities under the condition that all 36 elementary events are equiprobable.

1.4 Classical definition of probability. Combinatorics

29. A child plays with 10 letters of the alphabet: А, А, А, Е, И, К, М, М, Т, Т. What is the probability that with a random arrangement of the letters in a row he will obtain the word "МАТЕМАТИКА"?

30. In the elevator of an 8-story building, 5 persons entered on the first floor. Assume that each of them can, with equal probability, leave on any of the floors, starting with the second. Find the probability that all five will leave on different floors.

31. A cube, all sides of which are painted, is sawn into a thousand small cubes of the same dimensions. The small cubes obtained are carefully mixed. Determine the probability that a small cube selected at random will have two painted sides.

32. The same part can be made from material A or from material B. In order to decide which material endures the bigger load, n parts of each material were made and tested. Denote by $x_i(y_j)$ the limiting load which the i-th (j-th) part from the material A (B) endures. All the x_i and y_j obtained were distinct. It was decided to carry out the processing of the results of the experiments, using the Wilcoxon criterion.[1]) To this end, x_i,

---

[1]) See B. L. VAN DER WAERDEN. *Mathematical statistics.* New York, Springer-Verlag (1969).

y_j were arranged in a common series in the order of increasing magnitude, and for each j, there was found n_j, the number of x's occurring before y_j. It turned out that $\sum n_j \leqslant m$. On the basis of this the deduction was made that the parts made of material A were better. If the parts made of both materials are of the same quality, i.e., all arrangements of x's and y's in a series are equiprobable, find the probability of finding the inequality pointed out above for $n = 4$ and $m = 2$.

33. A deck of playing cards contains 52 cards, divided into 4 different suits with 13 cards in each suit. Assume that the deck is carefully shuffled so that all permutations are equiprobable. Draw 6 cards. Describe the space of elementary events.

a) Find the probability that among these cards there will be a king of diamonds.

b) Find the probability that among these cards there will be representatives of all suits.

c) What is the smallest number of cards one must take from the deck so that the probability that among them one encounters at least two cards of the same face value will be greater than $\frac{1}{2}$?

34. n friends sit down at random at a round table. Find the probability that:

a) two fixed persons A and B sit together with B to the left of A;

b) three fixed persons A, B and C sit together with A to the right of B and C to the left of B;

c) find these same probabilities in the case when the friends sit in a row on one side of a rectangular table.

35. Two numbers are chosen at random from the sequence of numbers $1, 2, \ldots, n$. What is the probability that one of them is less than k and the other is greater than k, where $1 < k < n$ is an arbitrary integer?

36. From the sequence of numbers $1, 2, \ldots, N$, n numbers are chosen at random and arranged in order of increasing magnitude: $x_1 < x_2 < \cdots < x_n$. What is the probability that $x_m \leqslant M$? Find the limit of this probability when $M, N \to \infty$ so that $M/N \to \alpha > 0$.

37. There are n tickets in a lottery, of which m are winners. How large is the probability of a win for a person holding k tickets?

38. In a lottery of forty thousand tickets, valuable winnings fall on three tickets. Determine:

a) the probability of obtaining at least one valuable prize if one has thousand tickets;

b) how many tickets is it necessary to acquire so that the probability of obtaining a valuable prize will not be less than 0.5?

39. In a sample made up of N parts, there are M defective ones. n parts are chosen at random from this sample $(n \leqslant N)$. What is the probability that among them there are m defective ones $(m \leqslant M)$?

40. Let $\varphi(n)$ denote the number of positive integers $\leqslant n$ and relatively prime to n. Prove that

$$\varphi(n) = n\Pi\left(1 - \frac{1}{p}\right),$$

where the product is taken over all prime numbers p which divide n. *Hint.* Consider the problem in which one number is chosen at random from the numbers $1, 2, ..., n$. Evaluate the probability that it will be relatively prime to n.

41. In a box there are n pairs of boots. $2r$ boots $(2r < n)$ are chosen at random from among them. What is the probability that, among the chosen boots,

a) paired boots are absent;

b) there is exactly one complete pair;

c) there are exactly two complete pairs?

42. A group consisting of $2N$ boys and $2N$ girls is divided in a random manner into two equal parts. Find the probability that the number of boys and girls is the same in each part. Calculate this probability using *Stirling's formula.*[1])

43*. In an urn there are n white and m black balls; $m < n$. Successively, without returning, all the balls are taken out. Let $M(k)$ be the number of black balls taken out after k steps, $N(k)$ the number of white balls taken out after k steps. Find the probability P that for all $k = 1, 2, ..., n+m$, $M(k) < N(k)$.

44.* *Banach's problem.* A certain mathematician carries two boxes of

---

[1]) Stirling's formula: $n! \sim \sqrt{2\pi}\, n^{n+\frac{1}{2}} e^{-n}$.

matches. Each time that he wants to get a match he chooses at random one of the boxes. Find the probability that when, for the first time, the mathematician finds one box empty, there are r matches in the other box ($r = 0, 1, 2, ..., n$; n is the number of matches which were originally in each of the boxes).

45. Each of n sticks is broken into two parts, a long and a short one. After that, the $2n$ parts obtained are collected into n pairs; from each pair a new "stick" is made. Find the probability that:
    a) the $2n$ parts are reassembled to form the original n sticks;
    b) each long part is joined to a short part.

46. Show that it is more probable to get at least one ace with four dice than at least one double ace in 24 throws of two dice. (This is known as *de Méré's paradox*. Chevalier de Méré, a gambler, thought that the two probabilities ought to be equal and blamed mathematics for his losses.)

47. Find the probability that in dealing a deck of 52 cards to four players the first of them obtains exactly n pairs "ace – king of one suit".

48. In certain rural communities of Russia there was at one time the following puzzle. A girl clutches in her hand six blades of grass so that the ends of blades of grass hang above and below; a playmate ties these blades of grass pairwise above and below separately. If all six blades of grass are tied to form a single ring, then this means that the girl will marry during the current year.
    a) Find the probability that the blades of grass, when tied at random, form a ring.
    b) Solve this same problem for the case of $2n$ blades of grass.

1.5 Simplest problems on arrangements

49. Let $f(x_1, ..., x_n)$ be an analytic function of n variables. How many different derivatives of the r-th order of it are there?

50. In how many different ways can one set out on n plates, r "éclair" pastries and s "napoleon" pastries?

51. Consider a mechanical system consisting of r indistinguishable parts.

Fundamental concepts

In statistical mechanics one usually subdivides the phase space into a large number n of small regions or cells so that each of the r particles falls into one of the cells. Thus, the state of the entire system is described as the distribution of r particles in n cells; consequently it is uniquely defined by the collection of numbers $0 \leqslant m_i \leqslant r$ $(i=1, 2, ..., n)$, where m_i is the number of particles in the i-th cell. Photons, atomic nuclei and atoms, containing an even number of elementary particles, are subject to *Bose-Einstein statistics*, in which only distinct distributions are considered and to each of these distinct distributions there is assigned an equal probability. Find this probability.

52. *Continuation.* Electrons, protons, neutrons are subject to *Fermi-Dirac statistics* in which it is assumed that:

a) not more than one particle can occur in one cell and

b) all distinct arrangements satisfying the first condition have equal probability. Find this probability in the case when there are r particles and n cells.

53. Suppose there are r particles and n cells, and that Bose-Einstein statistics hold (see Problem 51).

a) Prove that the probability of the presence in a fixed cell of exactly k particles equals

$$q_k = C^{r-k}_{n+r-k-2}/C^r_{n+r-1}.$$

b) Show that $q_0 > q_1 > q_2 > ...$

c) Prove that if n and r increase indefinitely where the mean number of particles, r/n, arriving at one cell, tends to $\lambda < \infty$, then $q_k \to (\lambda^k/(1+\lambda)^{k+1})$ (the right member is known by the name *geometric distribution*).

d) Prove that the probability that exactly m cells remain empty equals

$$\rho_m = (C^m_n \cdot C^{r-1}_{n-m-1})/C^r_{n+r-1}.$$

54. If, in a distribution of r particles in n cells, all n^r distributions have equal probability, then we speak of *Maxwell-Boltzmann statistics* (see also Problems 51 and 52). Find the probability that:

a) the first cell contains k_1 particles, the second k_2 particles, and so forth, where $k_1 + k_2 + \cdots + k_n = r$;

b) for $n = r$ none of the cells is empty;

c) for $n = r$ only one cell remains empty.

55.* A flow of k particles is caught by a system of n counters, registering the particles (Maxwell-Boltzmann statistics). Each particle falls into any of the counters with the same probability. What is the probability that the presence of particles will be noted by exactly r counters?

1.6   Geometric probability

56. A net of geographical coordinates is drawn on a sphere. The sphere is placed randomly on a plane. What is the probability that:

a) the point of contact with the plane lies between the 0-th and 90-th degrees east longitude on the sphere;

b) the point of contact with the plane lies between the 45-th and 90-th degrees north latitude on the sphere;

c) the minor arc of the great circle joining the point of tangency with the North Pole is less than α?

57. A point is chosen at random in the interior of a circle of radius R. The probability that the point falls inside a given region situated in the interior of the circle is proportional to the area of this region. Find the probability that:

a) the point occurs at a distance less than $r(r < R)$ from the center;

b) the smaller angle between a given direction and the line joining the point to the center does not exceed α.

58. On a circle of unit radius with center at the origin of coordinates, a point is selected at random. The probability that the chosen point lies on an arbitrarily given arc of the circle depends only on the length of this arc and is proportional to it. Find the probability that:

a) the projection of the chosen point on the diameter (axis of abscissas) occurs at a distance from the center not exceeding $r(r < 1)$;

b) the distance from the chosen point to the point with coordinates $(1, 0)$ does not exceed r.

59. A point M is chosen at random inside the square with vertices $(0, 0)$, $(0, 1)$, $(1, 0)$, $(1, 1)$. Let (ξ, η) be its coordinates. It is assumed that the

13

probability that the point falls inside a region situated entirely in the interior of the square depends only on the area of this region and is proportional to it.

a) Prove that for $0 \leqslant x, y \leqslant 1$,

$$P\{\xi < x; \eta < y\} = P\{\xi < x\} P\{\eta < y\} = xy.$$

b) For $0 < z < 1$, find

1) $P\{|\xi - \eta| < z\}$

2) $P\{\xi\eta < z\}$;

c) $P\{\min(\xi, \eta) < z\}$;
d) $P\{\max(\xi, \eta) < z\}$;
e) $P\{\frac{1}{2}(\xi + \eta) < z\}$.

60. On the plane, parallel lines are drawn at a distance $2a$ apart. A coin of radius $r < a$ is thrown at random on the plane. What is the probability that the coin does not intersect any of the (infinitely many) lines?

61. On an infinite chess board with side of a square equal to a, a coin of diameter $2r < a$ is thrown at random. Find the probability that:
a) the coin falls entirely in the interior of one of the squares;
b) the coin intersects no more than one side of the square.

62. Let ξ, η be defined as in Problem 59. Find the probability that the roots of the equation

$$x^2 + \xi x + \eta = 0$$

a) are real;
b) both positive.

63. A point M is chosen at random inside a triangle ABC in which AB has length l, BC has length k and the angle ABC is a right angle. Find the joint distribution of the length h of the perpendicular drawn from the point M onto AB, and the angle $\alpha = \angle MAB$ (i.e., for all x and y find the probability that the events $\{h < x\}$ and $\{\alpha < y\}$ are realized simultaneously).

64. On a horizontal plane foil there exists a point source of radioactivity sending rays uniformly in all directions of space above the foil. If a screen is set up parallel to the foil plane at a unit distance from it, then on this screen one can observe point charges registered by the radioactivity. Find the probability that a particular charge will occur in the part of the screen situated in the interior of the circle of radius R with center located over the source of radioactivity.

65. Let ξ and η be defined as in Problem 59 and let $\rho^2 = \xi^2 + \eta^2$, and $\phi = \arctan \eta/\xi$. Find the joint distribution of ρ and ϕ, i.e., for all x and y find the probability $P\{\{\rho < x\} \cap \{\phi < y\}\}$.

1.7 Metrization and ordering of sets

66. Show that $\rho(A, B) = P\{A \triangle B\}$ satisfies all the axioms of a metric space,[1]) except the axiom $\rho(A, B) = 0$ if and only if $A = B$; in other words, show that for arbitrary events A, B, C, we always have $\rho(A, B) + \rho(B, C) \geqslant \geqslant \rho(A, C) \geqslant 0$.

67. We agree to say that the property b_{ijk} holds for the events (sets) A_i, A_j, A_k if the following two conditions are satisfied:
  1) $A_i \cap \bar{A}_j \cap A_k = \emptyset$;
  2) $\bar{A}_i \cap A_j \cap \bar{A}_k = \emptyset$.
Prove that:
  a) if the property b_{ijk} holds, then

$$\rho(A_i, A_j) + \rho(A_j, A_k) = \rho(A_i, A_k);$$

  b) if $P\{A\} = 0$ implies $A = \emptyset$, then the converse assertion is also true.
68. Show that b_{ijm} does not always follow from the properties b_{ijk} and b_{jkm} (see the preceding problem).

69. Let $A^* = \{A_1, ..., A_n\}$ and $B^* = \{B_1, ..., B_n\}$ be two families of nested sets, $A_{j+1} \supseteq A_j$ and $B_{j+1} \supseteq B_j (j = 1, 2, ..., n-1)$, where $A_n \cap B_n = \phi$. Let C be a set such that $A_n \cap C = B_n \cap C = \emptyset$. Then the sequence of sets

$$L = (L_1, ..., L_n),$$

where

$$L_i = A_i \cup B_{n-i+1} \cup C$$

---

[1]) See P. S. ALEKSANDROV. *Introduction to the general theory of sets and functions.* Gostekhizdat, M. (1948) p. 227.

will be said to be *linearly ordered*. Prove that for $i \leqslant j \leqslant k$, L_i, L_j, L_k are b_{ijk}. From this, by virtue of Problem 67, it follows that in this case

$$\rho(L_i \cap L_j) + \rho(L_j \cap L_k) = \rho(L_i \cap L_k).$$

70*. *Continuation.* Prove that if $R^* = \{R_1, ..., R_n\}$ is a sequence of sets such that for all $i, j, k = 1, 2, ..., n$ and $i \leqslant j \leqslant k$ for R_i, R_j, R_k there holds the property b_{ijk}, then R^* is a linearly ordered sequence of sets.

See page 125 for the answers on problems 1–69.

2 *Application of the basic formulas*

The material of this chapter corresponds basically to §§9, 10 of the text-book by B. V. GNEDENKO. The central problems here are those involving application of the total probability formulas. To solve these problems, one needs to know how to break down a complicated problem into a series of simpler ones.

The formulation of the problems presupposes that the reader knows the definitions of *conditional probability* and *independence* and the *formula for total probability*.

If $P\{B\} > 0$, then the conditional probability of the event A under the condition B, $P\{A \mid B\}$, is defined by means of the formula

$$P\{A \mid B\} = \frac{P\{A \cap B\}}{P\{B\}}.$$

In practice, this formula is usually used to calculate $P(A \cap B)$.

Independence. The events A_1, ..., A_n are called *independent* if, for an arbitrary $1 \leqslant r \leqslant n$ and arbitrary $1 \leqslant i_1 \leqslant i_2 \leqslant \cdots \leqslant i_r \leqslant n$, the relation

$$P\{A_{i_1} \cap A_{i_2} \cap \ldots \cap A_{i_r}\} = \prod_{k=1}^{r} P\{A_{i_k}\}$$

holds. The random variables[1] ξ_1, ξ_2, ..., ξ_n are called *independent* if, for arbitrary x_1, x_2 ..., x_n, the equation

$$P\left\{\bigcap_{i=1}^{n} (\xi_i < x_i)\right\} = \prod_{i=1}^{n} P\{\xi_i < x_i\}$$

holds.

Formula for total probability. If the events $B_i (i = 1, 2, ..., n)$ are such that $B_1 \cap B_j = \emptyset \, (i \neq j)$, $\bigcup_{i=1}^{n} B_i = E$ (i.e., the B_i are pairwise disjoint and

---

[1] See page 35.

their union yields the entire space of elementary events) and $P\{B_i\}>0$, then the formula

$$P\{A\} = \sum_{i=1}^{n} P\{A \cap B_i\} = \sum_{i=1}^{n} P\{B_i\} \, P\{A \mid B_i\}$$

holds.

Discrete distributions

1. *Binomial law.* The number of successes in n independent Bernoulli trials, each with probability p of success, has the binomial distribution:

$$P\{k \text{ successes}\} = \binom{n}{k} p^k q^{n-k} \quad \text{where} \quad q = 1 - p.$$

2. *Multinomial law.* Suppose that each of n independent trials can result in r mutually exclusive and exhaustive outcomes $a_k \, (1 \leqslant k \leqslant r)$, and that the probabilities $P_k = P\{a_k\}$ are the same at each trial. Then

$$P\{\text{outcome } a_i \text{ occurs } t_i \text{ times, } 1 \leqslant i \leqslant r\} = \frac{n!}{t_1! \, t_2! \ldots t_n!} \, P_1^{t_1} P_2^{t_2} \ldots P_r^{t_r}$$

provided $t_1 + t_2 + \cdots + t_r = n$. We express this by saying that the joint distribution of the numbers of occurrences of the a_k is multinomial.

3. *Geometric law.* The probability P obeys the geometric law with parameter p if $P\{x\} = p(1-p)^{x-1}$, $x = 1, 2, \ldots$, and 0 otherwise.

4. *Hypergeometric law.* The probability P obeys the hypergeometric law with parameters N, n and p, $[N=1, 2, \ldots, n \, \varepsilon \, \{1, 2, \ldots, N\}$, and $p=0, 1/N, 2/N, \ldots, 1]$ if

$$P\{x\} = \frac{\binom{Np}{x} \binom{N(1-p)}{n-x}}{\binom{N}{n}}, \qquad x = 0, 1, \ldots, n, \quad 0 \text{ otherwise}.$$

5. *Poisson law.* The probability P obeys the Poisson law with parameter $\lambda > 0$ if

$$P\{x\} = e^{-\lambda} \frac{\lambda^x}{x!}, \qquad x = 0, 1, 2, \ldots, 0 \text{ otherwise}.$$

Continuous distributions

1. *Uniform law.* The uniform law over the interval α to β, $\alpha < \beta$, is specified by the density function[1]

$$f(x) = \begin{cases} \dfrac{1}{\beta - \alpha} & \alpha < x < \beta \\ 0 & \text{otherwise}. \end{cases}$$

2. *Normal law.* The normal law with parameters μ and σ, $-\infty < \mu < \infty$, $\sigma > 0$, is given by the density function

$$f(x) = \frac{1}{\sigma\sqrt{2\pi}} e^{-\frac{1}{2}((x-\mu)/\sigma)^2}, \qquad -\infty < x < \infty.$$

3. *Exponential law.* The exponential law with parameter $\lambda > 0$ is given by the density function

$$f(x) = \begin{cases} \lambda e^{-\lambda x} & x \geq 0 \\ 0 & x < 0. \end{cases}$$

4. *The χ^2 distribution.* The χ^2 distribution with parameters $n = 1, 2, \ldots$ and $\sigma > 0$ is given by the density function

$$f(x) = \begin{cases} \dfrac{1}{2^{n/2}\sigma^n \Gamma(n/2)} x^{(n/2)-1} e^{-(x/2\sigma^2)} & x > 0 \\ 0 & x \leq 0. \end{cases}$$

$\chi^2(n, 1)$ is called the χ^2 distribution with n degrees of freedom.

$$\left[\Gamma(t) = \int\limits_0^\infty x^{t-1} e^{-x} \, dx \right].$$

By the statement "$f(x) = O(g(x))$ as $x \to c$", we mean that the ratio $f(x)/g(x)$ remains bounded as $x \to c$. By the statement "$f(x) = o(g(x))$ as $x \to c$" we mean that the ratio $f(x)/g(x)$ tends to zero as $x \to c$. More formally, we mean that to each $\varepsilon > 0$, there corresponds a $\delta(\varepsilon) > 0$ such that $|f(x)| \leq \varepsilon |g(x)|$ whenever $c - \delta(\varepsilon) < x < c + \delta(\varepsilon)$.

For additional background one may consult [2], [3] and [11].

---

[1] See page 36.

2.1 Conditional probability. Independence

71. A student came to an examination knowing only 20 of 25 questions of the program. The examiner gave the student 3 questions. Using the concept of conditional probability, find the probability that the student knows all of these questions. Find the same probability using the classical definition of probability.

72. According to information about the consumption of repair parts, it was established that, in the repair of automobile engines, part No. 1 was changed in 36%, and part No. 2 in 42%, of the cases examined, and that both of these parts were changed simultaneously in 30% of the cases. On the basis of these data, is it possible to deduce that replacement of part No. 1 and replacement of part No. 2 are statistically connected with one another? Find the probability that in repairing an engine, part No. 2 will be changed, under the condition that part No. 1 has been changed.

73. Investigate the connection between dark-colored eyes of the father (event A) and of the son (event B) on the basis of the following data obtained in a population census in England and Wales in 1891. Dark-eyed fathers and dark-eyed sons ($A \cap B$) comprised 5% of all those observed, dark-eyed fathers and light-eyed sons ($A \cap \bar{B}$), 7.9%, light-eyed fathers and dark-eyed sons ($\bar{A} \cap B$), 8.9%, light-eyed fathers and light-eyed sons ($\bar{A} \cap \bar{B}$), 78.2%.

74. An electric circuit between the points A and B is made up according to the scheme shown in Fig. 1. Various elements of the circuit go out of commission independently of one another. The probability of elements of the circuit going out of commission during the time T are the following:

Element	K_1	K_2	L_1	L_2	L_3
Probability	0.1	0.2	0.4	0.7	0.5

Determine the probability of a current break during the indicated interval of time.

75. In searching for a certain book, a student decided to try three libraries. The libraries are stocked independently of one another and, for each library, the following statement is true: there is a 50% probability that the library possesses the book and, if it does possess it, there is a 50% probability that the book has been borrowed by another reader. Find the probability that the student succeeds in obtaining the book.

20

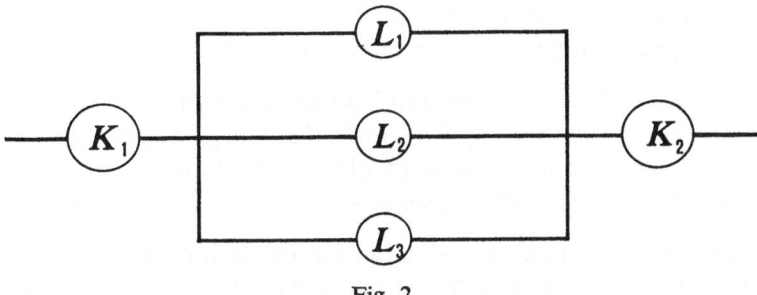

Fig. 2.

76. A marksman A hits the target under certain conditions of firing with probability $p_1 = 0.6$, marksman B with probability $p_2 = 0.5$ and marksman C with probability $p_3 = 0.4$. Each man fired one shot at the target and two bullets hit the bullseye. What is more probable: C hit the target, or not?

77. It is known that 5% of all men and 0.25% of all women are color-blind. A person chosen by chance suffers from colorblindness. What is the probability that this is a man? (It is assumed that there is an equal number of men and women.)

78. At a factory where bolts are manufactured, machines A, B, C produce respectively 25%, 35% and 40% of all bolts. Of the output of these machines, defects constitute 5%, 4% and 2% respectively. A bolt selected at random from the production turned out to be defective. What is the probability that it was produced by machine A? machine B? machine C?

79. It is known[1] that the probability that two twins are of the same sex is ≈ 0.64 where in general the probability of the birth of a boy is ≈ 0.51. Find the probability that the second of the twins is a boy under the condition that the first of them is a boy.

80. The probability that a letter is in the writing table equals p; with equal probability it can be in any one of eight drawers of the table. We looked in 7 drawers and did not find the letter. What is the probability that the letter is in the eighth drawer?

81. Three dice are thrown. What is the probability that at least one of them comes up with one eye if different faces came up on all three dice?

---

[1] See É. BOREL. *Probability and confidence.* Fizmatgiz, Moscow (1961) p. 39.

82. It is known that upon throwing 10 dice at least one unit came up. What is the probability that two or more ones came up?

83. For the experiment described in Problem 26, find for all m and n the conditional probabilities $P(A_m \mid B_n)$ and $P(B_n \cap A_m)$. Explain whether or not the trials A and B are dependent. (The trials $\mathfrak{A}$ and $\mathfrak{B}$ are dependent if for at least one pair (m, n) the events A_m and B_m are dependent.)

84. There are N children's wooden blocks; on each of them there can be glued the letter A or the letter B, or both of these letters together. We shall say that event A occurred if a randomly chosen block has the latter A, and event B if the block has the letter B. Can pictures be glued on in such a way that events A and B are independent?

85. Suppose the random variables ξ, and η are independent and identically distributed, with $P\{\xi=1\}=p>0$, $P\{\xi=0\}=1-p>0$. We introduce a new random variable, setting $\zeta=0$, if $\xi+\eta$ is an even number and $\zeta=1$ if $\xi+\eta$ is an odd number. For what values of p are the random variables ξ and ζ independent?

86. a) Prove that if $P(A)=0.9$, $P(B)=0.8$, then $P(A \mid B) \geqslant 0.875$.
   b) Prove that $P(A_2 \mid A_1) \geqslant 1 - P(\bar{A}_2)/P(A_1)$.

87. Let $P(A)=p$, $P(B)=1-\varepsilon$, where ε is small; estimate $P(A \mid B)$ from above and from below.

88. Construct an example showing that $P(A \cap B \cap C)=P(A)P(B)P(C)$ and $P(C)>0$ do not imply that $P(A)P(B)$.

89. Show that the pairwise independence of the events A, B, C does not imply their simultaneous independence.

90. It is known that the events A and B are independent and do not intersect. Find $\min(P(A),P(B))$.

91. Suppose given three pairwise independent events, all three of which cannot simultaneously occur. Assuming that they all have the same probability x, determine the largest possible value of x.

92. Given $P(A)$, $P(B)$, $P(C)$, $P(A \cap B)$, $P(A \cap C)$, find $P(B \cap C)$, $P(A \cap B \cap C)$, and $P(C \mid \bar{A} \cap \bar{B})$.

2.2 Discrete distributions: binomial, multinomial, geometric, hypergeometric

93. Two play at a game, tossing alternately a coin. The winner is counted to be the one who first comes up with a tail. Describe the space of elementary events. Find p_k, the probability that the game will terminate with the k-th toss. How many times greater is the probability of winning for the beginner?

94. In a sample of cotton, there are about 20% short fibers. What is the probability of not finding a single short fiber in a random selection of n fibers?

95. For weaving, white and colored cotton are mixed equally. What is the probability that among 5 randomly selected fibers of the mixture one finds less than 2 colored fibers?

96. Two equally skilled marksmen alternately shoot at a target. Each is allowed to have at most two shots. The first one who hits the target obtains a prize.

a) If the probability of hitting the target with a single shot is $p = \frac{1}{5}$, then, what is the probability that the marksmen obtain a prize?

b) Compare the chances of winning of the first and second to shoot. What if the number of shots is not restricted?

97. What is more probable, to win with an equally strong opponent:
   a) 3 games of 4 or 5 of 8;
   b) not less than 3 games of 4 or not less than 5 of 8;
   c) not more than n of $2n$ games or more than n of $2n$ games;
   d) not more than n of $2n + 1$ games or more than n of $2n + 1$ games?

98. The John Smith problem. In 1693 John Smith posed the following question: are the chances for success the same for three persons if the first one must obtain at least one "six" in throwing a die 6 times, the second, not less than two-sixes with 12 throws, and the third, not less than three sixes with 18 throws? The problem was solved by Newton and Tollet who showed that the first person has more chance of winning than the second and the second more than the third. Obtain this result.

99. Assume that the die has s faces, $s \geqslant 2$; that each of them comes up with equal probability. Denote by $g(t, n)$ the probability that with t throws of the die a given face will come up less than n times. Prove the following:

    a) $g(sn, n)$ decreases as s increases, with fixed n;

    b) $g(sn, n) < \frac{1}{2}$;

    c) $g(2n, n) \to \frac{1}{2}$ as $n \to \infty$.

100. In order to find out how many fish there are in a lake, 1000 fish are caught, marked and put back into the lake. For what number of fish in the lake will the probability be greatest of encountering 10 marked fish among 150 fish caught a second time?

101. Among the cocoons of a certain sample, 30% are colored. What is the probability that among 10 cocoons randomly selected from the sample, 3 will be colored? Not more than 3 will be colored?

102. A mechanical control checks parts, each of which (independently of the other parts) can turn out to be defective with probability p.

    a) What is the probability that of 10 checked parts only one turned out to be defective?

    b) Find the probability that the first defective part turns out to be the k-th part checked.

    c) Find the probability that 10 successive parts turn out to be good under the condition that the preceding $l = 5$ parts were also good. Does this probability depend on l?

    d) Find the distribution of the number of good parts found in checking between two successive defective parts.

103. Two persons play the following game. The first person writes down one of two numbers: zero or one, and the second person strives to guess which of the two numbers the first player wrote. The second player noticed that the first chooses the digits independently of one another and that, on each occasion, there is a probability $p = 0.6$ that he chooses zero. What must be the strategy of the second player, i.e., with what probability must he call each of the numbers in order to attain the largest number of correct guesses? Find the distribution of the number of correct guesses among two sequences of failures under the condition that the second player calls zero with probability $q = \frac{1}{2}$ independently of the results of the preceding guesses.

104. On a segment AB of length L there are located at random, independently of one another, five points. The probability of a point falling on some part of the segment depends only on the length of this part and is proportional to it. Find the probability that:

a) two points will occur at a distance less than b from the point A, and three at a distance greater than b;

b) two points will occur at a distance less than l from A, one at a distance r, with $l<r<b$, and two points at a distance greater than b.

105. A square is inscribed in a circle.

a) What is the probability that a point located at random in the interior of the circle turns out to be also interior to the square?

b) What is the probability that of 10 points located at random independently of each other in the interior of the circle, four fall into the square, three on one segment and one each on the remaining three segments?

106. The probability that a camouflaged opponent is in a shelled area equals 0.3; the probability of hitting in this case is equal to 0.2 for each individual shot. One hit is sufficient to destroy. What is the probability of destruction with 2 shots? What is the probability of destruction with 10 shots?

107. For a certain competition, each of the athletic leagues A and B picks three teams. A's first team plays B's first and wins with probability 0.8, A's second plays B's second and wins with probability 0.4 and A's third plays B's third and wins with probability 0.4. Find the probability that league A wins at least two games of the three.

108. Two chess players A and B agreed to play a match under the following conditions: for a victory, A must win 12 games before B wins 6; a tie is not counted. As A usually wins twice as frequently as B, the probability that A wins a particular game can be assumed to be equal to $\frac{2}{3}$. The game had to be terminated after A had won eight games and B four. Victory was decided credited to the one for whom the probability of final win was greater. Who was the victor?

109. Let A_k be the event that, in checking the first k parts coming up for control, not a single defective one is discovered. It is known that for arbitrary integers k and $l \geqslant 0$, $P\{A_{k+l} \mid A_k\} = P\{A_1\}$, where $P\{A_1\} = 1 - q$. Find $P\{A_k\}$. Also find the probability that the number of good parts found before the first defective part equals l (compare with Problem 102; see also Problem 117).

110. An urn contains two balls: one white and one black. Successive trials are performed with return of the ball taken out of the urn. The number of trials is infinite.

a) What is the probability of taking out in the final analysis a white ball if, after an unsuccessful attempt, n more black balls are added to the urn?

b) What is the probability of taking out in the final analysis two white balls in succession if, after each unsuccessful attempt, one more black ball is added to the urn?

c) What is the probability of taking out in the final analysis two white balls in succession if, after each unsuccessful attempt, two more black balls are added to the urn?

111. Applying probabilistic-theoretical considerations, verify the following identities, in which $N \geqslant m \geqslant 1$:

a) $1 + \dfrac{N-m}{N-1} + \dfrac{(N-m) \cdot (N-m-1)}{(N-1) \cdot (N-2)}$

$\qquad + \cdots + \dfrac{(N-m) \ldots 2 \cdot 1}{(N-1) \ldots (m+1) \, m} = \dfrac{N}{m}$;

b) $1 + \dfrac{N-m}{N} \cdot \dfrac{m+1}{m} + \dfrac{(N-m) \cdot (N-m-1)}{N^2} \cdot \dfrac{m+2}{m} +$

$\qquad + \cdots + \dfrac{(N-m) \ldots 2 \cdot 1}{N^{N-m}} \cdot \dfrac{N}{m} = \dfrac{N}{m}$;

c) $1 + \dfrac{N-m}{N+1} \cdot \dfrac{m+1}{m} + \dfrac{(N-m)^2}{(N+1)(N+2)} \cdot \dfrac{m+2}{m} +$

$\qquad + \dfrac{(N-m)^3}{(N+1) \cdot (N+2) \cdot (N+3)} \cdot \dfrac{m+3}{m} + \cdots = \dfrac{N}{m}$.

Find the corresponding schemes of drawing balls from an urn.

2.3   Continuous distributions

112. Kirchhoff investigated 60 spectral lines of radiation of iron and found that each of these lines lies within $\frac{1}{2}$ mm of some solar Fraunhofer line. Determine whether this is "conicidence", given that the mean distance between adjacent solar lines is around 2 mm.

Hint. In solving this problem, assume that 60 "iron lines" have been drawn at random independently of one another on a diagram of the solar spectrum, and estimate the probability that each of these lines turns out to be closer than $\frac{1}{2}$ mm to some solar line. (Also see Problem 60.)

113. What is the probability that a triangle can be constructed from three segments taken at random? The lengths of the segments are independent and each is uniformly distributed over $(0, l)$.

114. A twig of length l is broken randomly into two parts, after which the larger of the parts is broken in two at a point chosen at random. Find the probability that a triangle can be constructed from the parts obtained.

115. Two points are located randomly, independently of one another, on the segment $[0, 1]$. Find the probability that from the segments: from zero to the left point; from the left point to the right point; and from the right point to 1 it is possible:
   a) to construct a triangle;
   b) to construct an acute triangle.

116. Two persons agreed to meet in a definite place between 6 and 7 o'clock; the one arriving earlier is to wait for the other for up to one-quarter of an hour. Calculate the probability that the meeting will transpire, under the assumption that the times of arrival of the two persons at the meeting place are independent and each uniformly distributed between 6 and 7 o'clock.

117. Suppose that the event A_t consists of a molecule experiencing a collision at time $t \geqslant 0$ and not experiencing a collision up to the moment of time t. It is known that

$$P\{A_{t+u} \mid A_t\} = P\{A_t\}.$$

It is also known that $P\{A_1\} = e^{-\lambda}$. Find $P\{A_t\}$.

Application of the basic formulas

118. *The Buffon problem.* On a horizontal plane there are drawn parallel lines at a distance $2a$ from one another. On the plane there is thrown at random a thin needle whose length equals $2l$, where $l \leqslant a$. By "random" we understand the following: first, the center of the needle falls at random on a line perpendicular to the lines drawn, and, second, the angle ϕ formed by the needle and the lines drawn has a uniform distribution, where the position of the center and ϕ are independent. Find the probability that the needle will intersect some line. What is the probability that there will be 5 intersections in 10 throwings? What is the probability that there will be one intersection with 10 throwings?

119. In Problem 63, let $l = k$. Find, for all x,

$$P\left\{ h < x \mid \alpha \leqslant \frac{\pi}{6} \right\}.$$

120. For a ball-bearing assemblage it is necessary that among R – the radius of the outer ring, r – the radius of the inner ring, and d – the diameter of the balls (the balls may be assumed to be spherical) there exists the following relation:

$$0 \leqslant R - r - d \leqslant \delta.$$

Assume that R, r, d are independent and uniformly distributed on the segments [50.0, 51.0], [40.0, 41.0], [9.5, 10.0], respectively. Find the probability of the ball-bearing assemblage in the case $\delta = 0.5$ mm.

121. N points are scattered at random independently of one another in a sphere of radius R.

a) What is the probability that the distance from the center to the nearest point will not be less than r?

b) To what does the probability, found in (a) above, tend if

$$R \to \infty \quad \text{and} \quad \frac{N}{R^3} \to \tfrac{4}{3}\pi\lambda\,?$$

Remark. This problem is adapted from stellar astronomy: in a neighborhood of the sun, $\lambda = 0.0063$ if R is measured in parsecs.

2.4 Application of the formula for total probability

122. From among the 64 squares of a chess board, two different squares are chosen at random and two equal pieces of the white and black colors are placed on them. What is the probability that these pieces conquer one another if two rooks were placed? Two bishops? Two knights? Two queens?

123. From an urn containing 3 white and 2 black balls, there were transferred two balls, taken out at random, into an urn containing 4 white and 4 black balls. Find the probability of taking out a white ball from the second urn.

124. Three urns contain white and black balls. There are 2 white and 3 black balls in the first, 2 white and 2 black balls in the second, 3 white and 1 black ball in the third. A ball is transferred from the first urn into the second. After this a ball from the second urn is transferred to the third. Finally, a ball is transferred from the third urn to the first.

    a) What composition of the balls in the first urn is the most probable?

    b) Determine the probability that the composition of balls in all urns remains unmodified.

125. Someone (a certain Ivan!!) does not know all of the examination tickets. In what case will the probability of drawing an unknown ticket be the smallest for him: when he draws the ticket first or last?

126. The probability that the articles of some production process satisfy the standard equals 0.96. Assume that the system of inspection is simplified, yielding a positive result with probability 0.98 for articles which satisfy the standard and with probability of 0.05 for articles which do not satisfy the standard. What is the probability that an article that has passed the test satisfies the standard?

127. Assume that the probability of hitting the target with one shot equals p, and the probability of destroying the target with $k \geqslant 1$ strikes is $1 - q^k$. What is the probability that the target is destroyed if n shots were fired?

128. In a sample of N parts there are $M < N$ defective ones. From the sample there are selected at random $n < N$ parts which are subjected to a complete check. Errors are possible in the check; thus, with probability p a defective part is adjudged "suitable" and with probability q a good one is adjudged "defective". Find the probability that m parts will be adjudged "defective".

129. Let ξ be a nonnegative integer-valued random variable, taking the value $k = 0, 1, 2, \ldots$ with probability $(\lambda^k/k!\, e^{-\lambda})$. An experiment consists of choosing ξ points independently of one another on the segment [0.1]. Denote by x_i the number of points falling on the interval $((i-1)/n, i/n)$, $i = 1, 2, \ldots, n$. Prove that for $\lambda = n$ the x_i are independent.

130. Assume that a certain insect lays k eggs with probability $(\lambda^k/k!)\, e^{-\lambda}$, and that the probability of the evolution of an insect from an egg is p. Assuming the mutual independence of the evolution of the eggs, find the probability that an insect will have exactly l offspring.

131. In experiment $\mathfrak{A}$, M mutually exclusive outcomes A_m are possible, and in experiment $\mathscr{B}$, N mutually exclusive outcomes B_n are possible. Show that the conditional probability $P(B_n \mid A_m)$ can be expressed in terms of the probabilities $P\{A_m \mid B_n\}$ and $P\{B_n\}$ in the following way:

$$P\{B_n \mid A_m\} = \frac{P\{A_m \mid B_n\}\, P\{B_n\}}{\sum\limits_{k=1}^{N} P\{A_m \mid B_k\}\, P\{B_k\}}.$$

This relation is known as *Bayes' formula*.

132. From an urn in which there were $m \geqslant 3$ white balls and n black balls, one ball of unknown color was lost. In order to determine the composition of the balls in the urn, two balls were taken out of it at random. Find the probability that a white ball was lost if it is known that the balls taken out turned out to be white.
Hint. Use the formula of Problem 131.

133. Before certain experiments are performed, the probabilities of the mutually exclusive and exhaustive hypotheses $A_1, A_2, \ldots, A_k$ are considered to be $\alpha_1, \alpha_2, \ldots, \alpha_k$. According to hypothesis A_i, the probability of occurrence of the event B at any particular realization of the experiment is p_i. It is known that for n_1 independent trials, the event B occurred m_1 times. It is also known that in the subsequent series of n_2 trials, the event B

occurred m_2 times. Prove the following property of Bayes' formula: the *a posteriori* probabilities of the hypotheses, calculated after the second series of trials and taking into account the probabilities of these hypotheses after the first series of trials, are equal to the a posteriori probabilities, calculated simply on the basis of the series of $n_1 + n_2$ trials, in which the event B occurred $m_1 + m_2$ times.

134. On a communications channel, one of three sequences of letters can be transmitted: AAAA, BBBB, CCCC, where the *a priori* probabilities of the sequences are 0.3, 0.4, 0.3, respectively. It is known that the action of noise on the receiver decreases the probability of a correct reception of a transmitted letter to 0.6. The probability of the [incorrect] reception of a transmitted letter as either of the two other letters increases to 0.2. It is assumed that the letters are distorted independently of one another. Find the probability that the sequence AAAA was transmitted if ABCA is received on the receiver.

2.5 The probability of the sum of events

135. The events A_i $(i = 1, 2, ..., n)$ are independent and $P\{A_k\} = p_k$. Find the probability:

    a) of the occurrence of at least one of these events;

    b) the occurrence of only one of them.

136. Let $A_1, A_2, ..., A_n$ be random events. Prove the formulas:

a) $$P\left\{ \bigcup_{k=1}^{n} A_k \right\} = \sum_{k=1}^{n} P\{A_k\} - \sum_{i \leq i \leq j \leq n} P\{A_i \cap A_j\} +$$
$$+ \sum_{1 \leq i \leq j < k < n} P\{A_i \cap A_j \cap A_k\} + \cdots + (-1)^{n+1} P\left\{ \bigcap_{i=1}^{n} A_i \right\};$$

b) $$P\left\{ \bigcap_{k=1}^{n} A_k \right\} = \sum_{k=1}^{n} P\{A_k\} - \sum_{k=1}^{n-1} \sum_{j=k+1}^{n} P\{A_k \cup A_j\} +$$
$$+ \sum_{k=1}^{n-2} \sum_{j=k+1}^{n-1} \sum_{i=j+1}^{n} P\{A_k \cup A_i \cup A_j\} - \cdots + (-1)^{n-1} P\left\{ \sum_{k=1}^{n} A_k \right\}.$$

137. A schoolboy, wishing to play a trick on his friends, gathered all the caps in the cloakroom and then he hung them in a random order. What is the probability P_n that at least one cap landed in its previous place, if there were altogether n hooks in the cloak-room and n caps on them? Find $\lim_{n \to \infty} P_n$.

138. In an urn there are n tickets with numbers from 1 to n. The tickets are taken out randomly one at a time (without returning). What is the probability that:

a) for at least one drawing, the number of the selected ticket coincides with the number of the trial being performed;

b) If m tickets are drawn $(m < n)$, the numbers of the drawn tickets will go in increasing order?

139. One term of the expansion of a determinant of the n-th order is chosen at random. What is the probability P_n that it does not contain elements of the principal diagonal? Find $\lim_{n \to \infty} P_n$.

140. For an arbitrary $n \geqslant 3$ we set

$$A_1 \triangle A_2 \triangle ... \triangle A_n = \{A_1 \triangle A_2 \triangle ... \triangle A_{n-1}\} \triangle A_n.$$

In analogy with Problem 136, prove that

$$P\{A_1 \triangle A_2 \triangle ... \triangle A_n\} = \sum_{k=1}^{n} (-2)^{k-1} \sum_{v_1 < v_2 < ... < v_k} P\{A_{v_1} \cap ... \cap A_{v_k}\}.$$

Find the probability that in Problem 137 an odd number of caps fell into their original places.

2.6 Setting up equations with the aid of the formula for total probability

141. The probability that a molecule, having experienced at the moment $t = 0$ a collision with another molecule, and not having other collisions up to moment t, experiences a collision in the interval of time from t to $t + \triangle t$ equals $\lambda \triangle t + o(\triangle t)$. Find the probability that the time of free motion (i.e., the time between two successive collisions) will be greater than t.

142. Assuming that, in the multiplication of bacteria, the probability that a bacterium divides into two new bacteria in the interval of time $\triangle t$ equals $a \triangle t + o(\triangle t)$ and does not depend on the number of bacteria present, nor on the number of preceding divisions, find the probability that, there are i bacteria present at time t if there was one bacterium at time 0.

143. *Continuation.* Let us assume in addition that, independently of its own preceding history and of the total number of bacteria present, a bacterium alive at time t will perish in the time interval $(t, t+\triangle t)$ with probability $a\triangle t+o(\triangle t)$. Set up the differential equations which the probabilities $P_r(t)$, that there are r bacteria at time t, must satisfy.

144. n mechanisms are switched in an electric power transmission line. The probability that a mechanism requiring energy at time t terminates its requirement at the moment $t+\triangle t$, equals $\alpha\triangle t+o(\triangle t)$. If, at the moment, t, the mechanism does not require energy, then the probability that it will require it at the moment $t+\triangle t$ equals $\beta\triangle t+o(\triangle t)$, independently of the work of the other mechanisms. Set up the differential equations which $P_r(t)$, the probabilities that at the moment t, r mechanisms will require energy, satisfy. Find the stationary solution of these equations.

145. Two players A and B, having capital a and b, respectively, play at a game of chance consisting of separate plays. At each play, A and B have the same probability, $\frac{1}{2}$, of winning. After each play, the loser pays 1 ruble to the winner. The game continues until the bankruptcy of one of the players. Find the probability that the second player goes bankrupt.

146. Assume that in the preceding problem the player A wins with probability $p > \frac{1}{2}$ and loses with probability $q = 1-p$. In this case what will be the probability of bankruptcy of the second player?

147.* Find an integer β such that in throwing dice, the probability of the event A, that a series of three successive aces is encountered earlier than the first series of β consecutive non-aces, is approximately equal to one-half.
Hint. Introduce the conditional probabilities u and v of the event A under the conditions that the results of the first trial are respectively ace and non-ace. Using the formula for total probability, set up the equations relating u and v.

148.* Consider a sequence of independent trials each with three possible outcomes A, B, C of corresponding probabilities p, q and r $(p+q+r=1)$.

Application of the basic formulas

Find the probability that

a) a series of α consecutive A-outcomes occurs earlier than the first series of β consecutive B-outcomes;

b) a series of α consecutive A-outcomes occurs before one gets either β consecutive B-outcomes or γ consecutive C-outcomes.

See page 128 for the answers on problems 71–147.

3 *Random variables and their properties*

The material of this chapter corresponds basically to Chapters 4 and 5 of the textbook by B. V. GNEDENKO. Recall that by a random variable we understand a (measurable)[1]) function on the space of elementary events $\Omega = \{\omega\}$.

Example. A die is thrown. The elementary events (outcomes) $\omega_1, \omega_2, ..., \omega_6$ are the faces with one, two, ..., six eyes. Let ξ be the random variable taking on the value 0 if an even number of eyes come up and 1 otherwise.

Then

$$\xi(\omega_1) = \xi(\omega_3) = \xi(\omega_5) = 1,$$
$$\xi(\omega_2) = \xi(\omega_4) = \xi(\omega_6) = 0.$$

The equation

$$\eta = f(\xi_1, \xi_2, ..., \xi_n),$$

where the ξ_i are random variables, defined on the same space of elementary events $\Omega = \{\omega\}$, is understood in the sense that, for every ω,

$$\eta(\omega) = f(\xi_1(\omega), \xi_2(\omega), ..., \xi_n(\omega)).$$

Example. Given two random variables ξ_1 and ξ_2, each of which can take on the values 0 and 1, let $\eta = \min(\xi_1, \xi_2)$. In this case, it is convenient to introduce four elementary events $\omega_1, \omega_2, \omega_3, \omega_4$ such that

$$\xi_1(\omega_1) = \xi_1(\omega_2) = 0; \quad \xi_1(\omega_3) = \xi_1(\omega_4) = 1;$$
$$\xi_2(\omega_1) = \xi_2(\omega_3) = 0; \quad \xi_2(\omega_2) = \xi_2(\omega_4) = 1.$$

It is easy to see that $\eta(\omega_1) = \eta(\omega_2) = \eta(\omega_3) = 0$, and $\eta(\omega_4) = 1$. By the

---

[1]) See page 82.

probability of the event $\{\xi < x\}$ we mean the probability that at least one of the elementary events for which $\xi(\omega) < x$ will be realized. More generally: for an arbitrary (Borel-measurable) set A of values ξ,

$$P\{\xi \in A\} = P\{\omega : \xi(\omega) \in A\}.$$

Numerical properties of random functions. A distribution function (d.f.) $F_\xi(x)$ [or $F(x)$] of a random variable ξ is defined by

$$F_\xi(x) = P\{\xi < x\}.$$

It follows from this definition that $F(x)$ is a non-decreasing, left-continuous function, with $1 - F(x) + F(-x) \to 0 \, (x \to \infty)$. The last property is frequently used to find the unknown constants occurring in the definition of a d.f. In the case when the derivative $p(x)$ of $F(x)$ exists almost everywhere and

$$\int_{-\infty}^{\infty} p(x)\,dx = 1,$$

$p(x)$ is called the density of the random variable ξ. Clearly,

$$\int_{-\infty}^{x} p(x)\,dx = F(x).$$

The random variables $\{\xi_n\}$ are *identically* (equally) *distributed* if $F_{\xi_i} = = F_{\xi_j}$, ξ_i, $\xi_j \in \{\xi_n\}$.

In order to characterize a multidimensional joint distribution of several random variables $\xi_1, \ldots, \xi_n$, we sometimes introduce a multidimensional d.f. $F_{\xi_1, \ldots, \xi_n}(x_1, \ldots, x_n) = P\{\xi_1 < x_1, \xi_2 < x_2, \ldots, \xi_n < x_n\}$. If for all $x_1, \ldots, x_n$, the equation

$$F_{\xi_1, \ldots, \xi_m}(x_1, \ldots, x_n) = \prod_{i=1}^{n} F_{\xi_i}(x_i)$$

holds, we say that the random variables $\xi_1, \ldots, \xi_n$ are *independent*. In analogy with the one-dimensional case, we define the multidimensional density of the distribution, $p_{\xi_1\ldots\xi_n}(u_1, \ldots, u_n)$. We have

$$F_{\xi_1\ldots\xi_n}(x_1, \ldots, x_n) = \int_{-\infty}^{x} \ldots \int_{-\infty}^{x} p_{\xi_1\ldots\xi_n}(u_1 \ldots u_n)\, du_1 \ldots du_n.$$

The *mean value* or *mathematical expectation* of a random variable ξ is defined to be the integral $\int_{-\infty}^{\infty} x\, dF_{\xi}(x)$, if it exists, i.e., if $\int_{-\infty}^{\infty} |x|\, dF_{\xi}(x) < \infty$. The expectation of ξ is usually denoted by $M[\xi]$ or $E[\xi]$. In calculating expectation, it is useful to understand its properties:

a) the expectation of a constant C equals C;

b) a constant can be brought out of the expectation sign, i.e.,

$$M[C\xi] = CM[\xi];$$

c) the expectation of a sum of random variables equals the sum of the corresponding expectations, i.e.,

$$M[(\xi + \eta)] = M[\xi] + M[\eta];$$

d) the expectation of the product of independent random variables equals the product of the corresponding expectations, i.e.,

$$M[\xi\eta] = M[\xi] \cdot M[\eta];$$

$M[\xi^k]$ is called the *k-th order moment of* ξ, and $M[\xi - M[\xi])^k]$ is called the *central moment of the k-th order of* ξ. $D[\xi] = M[\xi - M[\xi])^2]$ is also called the dispersion.[1]) The fundamental properties of dispersion are:

a) $D[C] = 0$;

b) $D[C\xi] = C^2 D[\xi]$;

c) if ξ and η are independent, then $D[\xi + \eta] = D[\xi] + D[\eta]$.

The *covariance* of ξ and η is defined by $\mathrm{Cov}[\xi, \eta] = M[(\xi - M[\xi])(\eta - M[\eta])]$.

One of the most useful characteristics of the measure of dependence, connecting two random variables ξ and η, is the *correlation coefficient*

$$\rho = \frac{M[(\xi - M[\xi])(\eta - M[\eta])]}{\sqrt{D[\xi]\,D[\eta]}}.$$

---

[1]) [Editor's note: dispersion is also commonly called "variance"; we also write $D[\xi] = \mathrm{Var}[\xi] = \sigma^2(\xi)$.]

The fundamental properties of ρ are:

    a) $-1 \leqslant \rho \leqslant 1$;

    b) if ξ and η are independent, then $\rho = 0$;

    c) $|\rho| = 1$ if, and only if, one of the random variables is a linear transformation of the other.

The random variables ξ and η are said to be *uncorrelated* if $M[\xi \cdot \eta] = M[\xi]M[\eta]$.

If $P\{A\} > 0$, then we can define $F_\xi(x \mid A) = P\{\xi < x \mid A\}$. In this case, by $M[x \mid A]$ we mean $\int_{-\infty}^{\infty} (x \, dF_\xi(x \mid A))$. For the definition of conditional mathematical expectation in the case when $P\{A\} = 0$, see Chapter 6. But we note here that in calculating a d.f. and moments it is convenient to use the generating and characteristic functions (see Problem 298, Chapter 5)[1].

Entropy and information. The *entropy* $H(\xi)$ of a discrete random variable ξ with a distribution defined by the sequence $\{p_i\}$ ($i = 1, 2, \ldots$), where $p_i = P\{\xi = x_i\}$, is defined to be

$$- \sum_i p_i \log_a p_i.$$

Entropy can be considered to be a measure of indeterminacy [see Problem 212].

Entropy is measured in units corresponding to the logarithm base a. In this problem book, we take $a = 2$. Suppose given two random variables ξ and η, and let $p_{ij} = P\{\xi = x_i; \ \eta = y_j\}$; then we define $H(\xi, \eta)$, the entropy of the random variables ξ and η, as

$$- \sum_{i,j} p_{ij} \log_a p_{ij}.$$

For every j, for which

$$p_{\cdot j} = \sum_i p_{ij} > 0,$$

we can define the conditional probability

$$p\{i \mid j\} = P\{\xi = x \mid \eta = y_j\} = \frac{p_{ij}}{p_{\cdot j}}$$

and the conditional entropy

$$H(\xi \mid \eta = y_j) = - \sum_i p(i \mid j) \log_a p\{i \mid j\}.$$

---

[1] See also page 71.

The quantity

$$H(\xi, \eta) = \sum p_{\cdot j} H(\xi \mid \eta = y_j)$$

is called the *mean conditional entropy of ξ relative to η*, and the quantity $I_\eta(\xi) = H(\xi) - H(\xi \mid \eta)$ is called the quantity of information, contained in η relative to ξ. It is not difficult to show that $I_\eta(\xi) = I_\xi(\eta) \geqslant 0$ (see Problem 217). The quantity of information is also denoted by $I(\xi, \eta)$. Problems 216–221 and 259–260 of Chapter 4 point out the suitability of using the concepts of entropy and information in the statistical theory of communications. In Problems 398–400 of Chapter 7 the concept of entropy is introduced for continuous distributions.

For additional reading see [3], [8], [10] and [12].

3.1 Calculation of mathematical expectations and dispersion

149. Find the d.f. and the mean value of the number of tosses of a coin in Problem 93.

150. The random variables ξ and η are independent, where $M[\xi] = 2$, $D[\xi] = 1$, $M[\eta] = 1$, $D[\eta] = 4$. Find the expectation and dispersion of:
   a) $\zeta_1 = \xi - 2\eta$;
   b) $\zeta_2 = 2\xi - \eta$.

151. Assume that in a lake there were 15,000 fish, 1000 of them marked [with radioactive tracers] (see Problem 100). 150 fish were fished out of the lake. Find the expectation of the number of marked fish among the fish caught.

152. Find the expectation and the dispersion of the number of short fibers among the randomly selected fibers in Problem 94.

153. In throwing n dice, determine the expectation, the dispersion, and central moment of the 3-rd order of the sum of the eyes on all dice.

154. Find the expectation and dispersion of the magnitude of the free motion of the molecule described in Problem 141.

155. The owner of a railway season ticket usually departs from home between 7:30 and 8:00 a.m.; the journey to the station takes from 20 to 30 min. It is assumed that the time of departure and duration of the journey are independent random variables, uniformly distributed in the corresponding intervals. There are two trains which he can ride: the first departs at 8:05 a.m. and takes 35 min.; the second departs at 8:25 a.m. and takes 30 min. Assuming that he departs on one of these trains, determine the mean time of his arrival at his destination. What is the probability that he will miss both trains?

156. Find the expectation and dispersion of the number of defective parts among n parts subjected to control; see Problem 102. Find also the expected number of good parts occurring between two successive defective ones.

157. In Problem 130, find the expectation of the number of descendants of the insect.

158. Two dice are thrown. Find the expectation of the sum of the scores if it is known that different sides came up.

159. The diameter of a circle is measured approximately. Assuming that its magnitude is uniformly distributed in the segment $[a, b]$, find the distribution of the area of the circle, its mean value and dispersion.

160. The density of the distribution of the absolute value of the velocity of motion of a molecule has the form

$$p(s) = 4 \sqrt{\frac{a^3}{\pi}} s^2 e^{-as^2}$$

(the constant a can be determined by the temperature of the gas and the mass of the particle observed: $a = m/2kT$, where k is the Boltzmann constant).

    a) Find the mean value of the path traversed by the molecule in a unit of time (expected motion of the molecule).

    b) Find the mean value of the kinetic energy of the molecule (the so-called "mean energy" of the molecule).

161. It is known that the probability of the breakdown of an electronic tube in the next $\triangle$ days, having functioned x days, equals $0.003\triangle + o(\triangle)$ independently of the quantity x. After a year of work, the lamp is changed even if it has not gone out of commission. Find the mean time of functioning of the lamp.

162. A two-dimensional distribution of a pair of integral random variables ξ and η is defined by means of the table

$$P\{\xi=i, \eta=j\}.$$

$\eta=j$	$\xi=i$					
	0	1	2	3	4	5
0	0.01	0.05	0.12	0.02	0	0.01
1	0.02	0	0.01	0.05	0.02	0.02
2	0	0.05	0.1	0	0.3	0.05
3	0.01	0	0.02	0.01	0.03	0.1

Find:
 a) $P\{\xi=2 \mid \eta=3\}$;
 b) $M[\xi \mid \eta=1]$;
 c) $M[\xi+\eta]$;
 d) $M[\xi^2 \mid \eta\leqslant 1]$;
 e) $P\{\xi+\eta\leqslant 5 \mid \eta\leqslant 2\}$;
 f) $M[\xi\cdot\eta \mid \eta\leqslant 1]$.

163. Prove that if $\xi_1, \xi_2, \ldots, \xi_n$ are independent, positive and identically distributed then

$$M\left[\left(\frac{\xi_1 + \cdots + \xi_k}{\xi_1 + \cdots + \xi_n}\right)\right] = \frac{k}{n} \quad \text{if} \quad k < n.$$

164. The random variable ξ takes on positive integer values. Prove that:

 a) $M[\xi] = \sum_{m\geqslant 1} P\{\xi \geqslant m\}$;

 b) $D[\xi] = 2 \sum_{m\geqslant 1} mP\{\xi \geqslant m\} - M[\xi(M[\xi] + 1)]$.

165. Define $D[\xi \mid A] = M[(\xi - M[(\xi \mid A))^2] \mid A]$. Prove that

$$D[\xi \mid A] = M[(\xi - M[\xi]^2 \mid A] - [M[\xi \mid A] - [\xi M]]^2.$$

166. Assume that the random variable ξ coincides, with probability p_i, with the random variable ξ_i and let $M[\xi_i]=M_i$. Prove that

$$D[\xi] = \sum_k p_k D[\xi_k] + D[\mu],$$

where μ takes on the values M_i with probability p_i.

167. Find the mean value and dispersion of the product of two independent random variables ξ and η with uniform distribution laws: ξ in the interval $[0,1]$ and η in the interval $[1,3]$.

168. Prove that if ξ and η are independent, then

$$D[\xi \cdot \eta] = D[\xi] \cdot D[\eta] + (M[\xi])^2 D[\eta] + (M[\eta])^2 D[\xi],$$

i.e.

$$D[\xi \cdot \eta] \geqslant D[\xi] D[\eta].$$

169. Let $\xi_1, \xi_2, ..., \xi_{n+1}$ be a sequence of mutually independent identically distributed random variables, taking on the value 1 with probability p and the value 0 with probability $q=1-p$. Set $\eta_i=0$ if $\xi_i+\xi_{i+1}$ is an even number, and $\eta_i=1$ if $\xi_i+\xi_{i+1}=1$. Find the expectation and the dispersion of

$$\zeta = \sum_{i=1}^{n} \eta$$

170. A large number N of people are subjected to a blood investigation. This investigation can be organized in two ways. 1. The blood of each person is investigated separately. In this case N analyses are needed.
2. The blood of k persons is mixed and the mixture obtained is analyzed. If the result of the analysis is negative, then this single analysis is sufficient for k persons. But if it is positive, then the blood of each one must be subsequently investigated separately, and *in toto* for k persons, $k+1$ analysis are needed. It is assumed that the probability of a positive result is the same for all persons and that the results of the analyses are independent in the probabilistic sense.

a) What is the probability that the analysis of the mixed blood of k persons is positive?

b) What is the expectation of the number of analyses ξ necessary in the second method of investigation?

c) For what k is the minimum expected number of necessary analyses[1]) attained?

171. A city consists of n apartments, where in each of n_j of them there live x_j inhabitants $(n_1 + n_2 + \cdots = n)$. Let

$$m = \sum_{j=1}^{n} \frac{n_j x_j}{n}$$

be the mean number of inhabitants per apartment. We also set

$$\sigma^2 = \sum_{j=1}^{m} \frac{n_j x_j^2}{n} - m^2 .$$

Random choice, without replacement, is made of r apartments and in each of these the number of inhabitants is counted. Let $X_1, ..., X_r$ be the resulting numbers. Prove that

$$M[X_1 + \cdots + X_r] = mr ; \qquad D[X_1 + \cdots + X_r] = \frac{\sigma^2 r (n - r)}{n - 1} .$$

(We remark that, for choice with replacement, the dispersion is greater.)

172. The number of inhabitants of a city is estimated by means of the following procedure of double choice. The city is subdivided into n regions. The number of apartments in the j-th region is known and is equal to n_j, so that $n = \sum n_j$ is the total number of apartments in the city. We denote by x_{jk} the unknown number of inhabitants in the k-th apartment of the j-th region (so that $x_j = \sum_k x_{jk}$ is the number of inhabitants in the j-th region, and $x = \sum x_j$ is the number of inhabitants in the city).

From the j-th region r_j living quarters are chosen and the number of people living in each of them is counted. Let X_{jk} be the number of inhabitants in the k-th living quarter among those chosen from the j-th region. Then $X_j = \sum X_{jk}$ is the total number of inhabitants in the living quarters chosen from the j-th region. Set

$$X = \sum \frac{n_j}{r_j} X_j .$$

---

[1]) As W. FELLER, from whose textbook this problem is adapted, points out, the second method gave a saving, in practice, in the number of analyses of up to 80%.

Show, using the result of the preceding problem, that

$$M[X] = x;$$

$$D[X] = \sum \sigma_j^2 n_j^2 (n_j - r_j) \frac{1}{r_j(n_j - 1)},$$

where

$$\sigma_j^2 = \frac{1}{n_j} \sum_k n_{jk} \left(x_{jk} - \frac{x_j}{n_j}\right)^2.$$

3.2 Distribution functions

173. Let $p(x)$ be the density of the random variable ξ. The constant C appears in its definition. Find it in the case when

a) $p(x) = \begin{cases} 0 & \text{for } x < 0; \\ Ce^{-x} & \text{for } x \geq 0; \end{cases}$

b) $p(x) = \begin{cases} 0 & \text{for } x < 0; \\ Cx^\alpha e^{-\beta x} & \text{for } x \geq 0 \, (\alpha > 0, \beta > 0); \end{cases}$

c) $p(x) = C(1 + x^2)^{-1}$.

174. For a given tramway line from the point 0 to L, the known function $F(a, b)$ represents the probability that a passenger riding on this line got on at a point $x < a$ and rides to a point $y \leq b$. It is required to determine:

a) the relative density of motion, namely the function $\phi(z)$ which represents the probability that a passenger riding on the given line, rides through the point z;

b) the probability $\phi_1(z)$ that he got on at the point z;

c) the probability $\phi_2(z)$ that he got off no later than z. Assuming that the functions introduced are continuous and differentiable, establish the dependence among them and the function $p(x, b)$ which expresses the probability density that a passenger who got on at the point x gets off at the point $b > x$.

175. A certain number of perfectly spherical balls, made from a homogeneous material, yield a symmetric distribution when grouped according to diameter. Show that if these balls are grouped by weight, the distribution will have a positive asymmetry (i.e., the third central moment will be positive).

176. Prove that an arbitrary distribution function possesses the following properties:

$$\lim_{x \to \infty} x \int_x^\infty \frac{1}{z} \, dF(z) = 0, \qquad \lim_{x \to 0+} x \int_x^\infty \frac{1}{z} \, dF(z) = 0.$$

177. Prove that if a random variable ξ has a moment of order k, then $\lim_{x \to \infty} x^k (1 - F(x) + F(-x)) = 0$.

178. Show that the sequence of moments of an arbitrary continuous distribution F is positive definite, i.e., for an arbitrary m and arbitrary real $x_1, x_2, \ldots, x_m$,

$$\sum_{i,k=0}^m \alpha_{i+k} x_i x_k > 0, \quad \text{where} \quad \alpha_l = \int_{-\infty}^\infty x^l \, dF(x).$$

3.3   Correlation coefficient

179. Let ξ and η be variables having finite moments of the second order. Show that $D[\xi + \eta] = D[\xi] + D[\eta]$ if, and only if, these variables are not correlated.

180. Prove that if the correlation coefficient ρ of two random variables ξ and η is such that $|\rho| = 1$, then there exist constants a and b such that $\xi = a\eta + b$.

181. Construct an example which shows that the correlation coefficient equal to zero does not imply that the corresponding random variables are independent.

182. The random variables $\xi_1, \xi_2, \ldots, \xi_n$ are independent and normally distributed, with parameters a, σ. Find the two-dimensional density of the distribution

$$\eta = \sum_{k=1}^m \xi_k \quad \text{and} \quad \zeta = \sum_{k=1}^n \xi_k \, (m < n).$$

45

183. The random variables ξ and η are independent and normally distributed, with the same parameters a and σ.

a) Find the correlation coefficient of the variables $\alpha\xi+\beta\eta$ and $\alpha\xi-\beta\eta$; also find their joint distribution.

b) Prove that $M[\max(\xi,\eta)]=a+\sigma/\sqrt{\pi}$.

184. The random vector (ξ,η) is normally distributed $M[\xi]=M[\eta]=0$; $D[\xi]=D[\eta]=1$. ρ is the correlation coefficient between ξ and η. Prove that

a) $\rho=\cos q\pi$, where $q=P\{\xi\eta<0\}$;

b) $M[\max(\xi,\eta)]=\sqrt{(1-\rho)/\pi}$;

c) the correlation coefficient of the variables ξ^2 and η^2 equals ρ^2.

185. Let $\xi_i\,(i=1,2,\dots,n)$ be independent and have the same distribution, with $M[(\xi-M[\xi]^3)]=0$. Prove that in this case the random variables

$$\bar{\xi}=\sum_{i=1}^{n}\xi_i \quad\text{and}\quad S^2=\sum_{i=1}^{n}(\xi_i-\bar{\xi})^2$$

are not correlated.

3.4 Chebyshev's inequality

186. Let ξ be a random variable having a finite dispersion. Prove Chebyshev's inequality, which states that

$$P\{|\xi-M[\xi]|\geqslant\varepsilon\}\leqslant\frac{D[\xi]}{\varepsilon^2}.$$

187*. Let ξ be an arbitrary random variable, where $M[\xi]=0$, $D[\xi]=\sigma^2$, and let $F(x)$ be the distribution function of ξ. Prove that, for

$$x<0,\ F(x)\leqslant\frac{\sigma^2}{\sigma^2+x^2},$$

and for

$$x>0,\ F(x)\geqslant\frac{x^2}{\sigma^2+x^2}.$$

Show by an example that these inequalities can turn into equalities for certain F's.

188. If we restrict ourselves to only certain classes of distributions, then sometimes one succeeds in sharpening the Chebyshev inequality. Thus, in 1821 Gauss showed that for unimodal distributions of continuous type, i.e., distributions whose density has a single maximum, and for an arbitrary $\varepsilon > 0$, we have that

$$P\{|\xi - x_0| \geqslant \varepsilon\tau\} \leqslant \frac{4}{9\varepsilon^2},$$

where x_0 is the mode, and $\tau^2 = D[\xi] + (x_0 - M[\xi])^2$ is the second moment with respect to the mode. If we use

$$S = \frac{M[\xi] - x_0}{\sqrt{D[\xi]}}$$

the measure of asymmetry introduced by K. Pearson, then from the given inequality we can obtain that for $\varepsilon > |s|$,

$$P\{|\xi - M[\xi]| \geqslant \varepsilon\sqrt{D[\xi]}\} \leqslant \frac{4(1 + s^2)}{9(\varepsilon - |s|)^2}.$$

Prove both of these inequalities.

Hint: First prove that if $g(x)$ does not increase for $x > 0$, then for an arbitrary $\varepsilon > 0$,

$$\varepsilon^2 \int\limits_{\varepsilon}^{\infty} g(x)\,dx \leqslant \frac{4}{9} \int\limits_{0}^{\infty} x^2 g(x)\,dx.$$

189. Generalization of the Chebyshev inequality.

a) Prove that if the random variable ξ is such that $M[e^{a\xi}]$ exists ($a > 0$ is a constant), then

$$P\{\xi \geqslant \varepsilon\} \leqslant \frac{M[e^{a\xi}]}{e^{a\varepsilon}}.$$

b) Let $f(x) > 0$ be a non-decreasing function. Prove that if $M[f(|\xi - M[\xi]|)]$ exists, then, for $\varepsilon > 0$,

$$P\{|\xi - M[\xi]| \geqslant \varepsilon\} \leqslant \frac{M[f(|\xi - M[\xi]|)]}{f(\varepsilon)}.$$

3.5 Distribution functions of random variables

190. ξ and η are independent, where $P\{\xi=0\}=P\{\xi=1\}=\frac{1}{2}$ and $P\{\eta<x\}=$ $=x\,(0<x<1)$. Find the distribution function

    a) $\zeta_1=\eta+\xi$;

    b) $\zeta_2=\eta+\frac{1}{2}\xi$;

    c) $\zeta_3=\xi\cdot\eta$.

191. Find the distribution function of the sum of the independent random variables ξ and η, the first of which is uniformly distributed in the interval $(-h, h)$ and the second has distribution function $F(x)$.

192. Concentration function. The quantity

$$Q_\xi(l) = \sup_x P\{x \leqslant \xi \leqslant x + l\}$$

is called the *concentration function* of the random variable ξ. Prove that for an arbitrary η independent of ξ, the concentration function of the sum $\xi+\eta$ is such that $Q_{\xi+\eta}(l)\leqslant Q_\xi(l)$.

193. Prove that if a random variable ξ has density $p_\xi(x)$, then for an arbitrary independent of ξ, the sum $\xi+\eta$ also has density $p_{\xi+\eta}(x)$, where $p_{\xi+\eta}(x)\leqslant\sup_x p_\xi(x)$.

194. Suppose the random variable ξ has distribution density $p(x)$. Find the distribution density of the random variable:

    a. $\eta=a\xi+b$, a and b are real numbers;

    b) $\eta=\xi^{-1}$;

    c) $\eta=\cos\xi$;

    d) $\eta=f(\eta)$, where $f(x)$ is a continuous monotone function.

195. Prove that, for an arbitrary random variable ξ with continuous distribution function $F(x)$, for $0\leqslant x\leqslant1$,

$$P\{F(\xi) < x\} = x.$$

196. A discrete random variable ξ has the Poisson distribution:

$$P\{\xi = m\} = \frac{\lambda^m e^{-\lambda}}{m!}, \quad m = 0, 1, 2, \ldots.$$

Let M be the mean of N independent realizations of ξ:

a) determine the mean and dispersion of M;

b) find the distribution of M;

c) construct the graph for the result of b) for $\lambda = 1$ with $N = 3$ and with $N = 10$.

197. A random variable ξ has the Cauchy density

$$p(x) = \frac{C}{1 + x^2}.$$

Let M be the mean of N independent realizations of ξ:

a) find C;

b) find the density M;

c) find the probability that each of two independent realizations of ξ is in absolute value less than unity.

198. Let $p_\xi(x)$, $p_\eta(y)$, $p_{\xi+\eta}(z)$ be the densities of the random variables ξ, η, $\xi+\eta$. Prove that if ξ and η are independent, then

$$p_{\xi+\eta}(z) = \int_{-\infty}^{\infty} p_\xi(z - y)\, p_\eta(y)\, \mathrm{d}y = \int_{-\infty}^{\infty} p_\xi(x)\, p_\eta(z - x)\, \mathrm{d}x.$$

199. The densities of the independent random variables ξ and η are equal to:

a) $p_\xi(x) = p_\eta(x) = \begin{cases} 0 & x \leqslant 0, \\ a\,e^{-ax} & x > 0; \end{cases}$

b) $p_\xi(x) = p_\eta(x) = \begin{cases} 0 & x \leqslant 0,\ x \geqslant a, \\ \dfrac{1}{a} & 0 < x < a; \end{cases}$

c) $p_\xi(x) = p_\eta(x) = \dfrac{1}{\sqrt{2}}\,e^{-x^2/2}.$

Find the distribution density of $\zeta = \xi/\eta$.

200. Find the distribution function of the product of the independent variables ξ and η, given their distribution functions $F_1(x)$ and $F_2(x)$, respectively.

201. The random variables ξ_1, ξ_2, ..., ξ_n, ... are independent and uniformly distributed in [0,1]. Let v be a random variable which is equal to that k for which the sum

$$S_k = \xi_1 + \xi_2 + \cdots + \xi_k$$

first exceeds 1. Prove that $M[v] = e$.

202. Let $\{\xi_i\}$ be a given sequence of independent random variables which take on the values 0 and 1 with the probabilities $\frac{1}{2}$. Find the distribution of the random variable

$$x = \sum_{i=1}^{\infty} \frac{\xi_i}{2^i}.$$

203.* In carrying out calculations by the Monte-Carlo method, frequently a sequence of independent normally distributed random variables is required. In an electronic computer, number-theoretic methods may be used to produce a sequence of independent random variables $\xi_1, \ldots, \xi_n, \ldots$ which are uniformly distributed on [0,1]. It turns out that there exists a function $\Psi(x)$ such that $\eta_i = \Psi(\xi_i)$ has a normal distribution. However, it is inconvenient to construct a sequence of normally distributed variables with the aid of Ψ since tabulation of Ψ requires too many memory cells. Usually, to construct normal variables, we proceed in the following way. We decompose the sequence ξ_i into pairs. And for each pair ξ_i, ξ_{i+1}, with the aid of the transformations

$$\phi = 2\pi\xi_i; \quad z = -\ln \xi_{i+1}; \quad r = \sqrt{2z};$$
$$\eta_i = r \cos \phi; \quad \eta_{i+1} = r \sin \phi$$

we obtain a sequence of independent normally distributed quantities. In connection with this, the following problems arise:
 a) find the function $\Psi(x)$;
 b) prove that z has an exponential distribution;
 c) prove that η_i and η_{i+1} are independent and have a normal distribution with parameters [0,1].

204.* n points are located on the segment [0,1]. Assuming that the points are located at random (i.e., each of them is situated independently of the others and distributed uniformly in [0,1]), find:

a) the density of the distribution $\zeta_1 = \max(\xi_1, \xi_2, ..., \xi_n)$;

b) the density of the distribution of the k-th point from the left;

c) the joint density of the distribution of the abscissas of the k-th and m-th points from the left $(k < m)$;

d) the density of the distribution $\zeta_2 = \max_i \xi_i - \min_i \xi_i$.

205.* Let $\xi_i (i = 1, ..., n)$ be independent identically distributed random variables. It is known that they are uniformly distributed on some segment where the segment itself is unknown. It is possible to construct several estimates of the center of this segment. For example, let

$$a_1 = \frac{1}{n} \sum_{i=1}^{n} \xi_i \quad \text{and} \quad a_2 = \tfrac{1}{2}\Big(\max_{1 \leqslant i \leqslant n} (\xi_i) + \min_{1 \leqslant i \leqslant n} (\xi_i) \Big);$$

prove that these estimates are unbiased, i.e., $M[a_1] = M[a_2] = a$, where a is the center of the segment. Prove that $D[a_1] > D[a_2]$, i.e., that the estimate a_2 is more effective than the estimate a_1.

206. Suppose the random variables $\xi_i (i = 1, 2, ..., n)$ are independent and identically distributed according to the law

$$F(x) = \begin{cases} 0 & \text{for} \quad x \leqslant 0, \\ 1 - e^{-x} & \text{for} \quad x > 0. \end{cases}$$

Find the distribution of $\zeta_n = \max_{1 \leqslant i \leqslant n}(\xi_i)$ and prove that it is possible to choose constants a_n such that the distribution of $\zeta_n - a_n$ tends to a limiting distribution law.

207.* Let $\xi_i (i = 1, ..., n)$ be independent and identically distributed with continuous distribution function $F(x)$. Denote the number of ξ_i which are less than x by $v(x)$. Prove that the distribution $D_n = \sup_x |v(x)/n - F(x)|$ does not depend on $F(x)$. Find the distribution of $v(x)$.

Remark. The fact that the distribution D_n does not depend on F plays an important role in statistics. The Kolmogorov criterion for the deviation of the empirical distribution function from the theoretical distribution function is based on it.[1]

---

[1] See B. L. VAN DER WAERDEN. *Mathematical statistics*. New York. Springer-Verlag (1969).

3.6   Entropy and information

208. Let $f(x)$ be a differentiable function defined on the segment $[0, l]$ and let $f(0)=0$ and $|f'(x)|<d$. How many binary units of information are necessary in order to determine, to within $\varepsilon>0$, the value of $f(x)$ at every point of the segment $[0, l]$?

209. There are n coins. All of them look alike; however, one of them is counterfeit. It is known that the counterfeit coin is heavier than the others. There are also scales with two pans. There is no set of weights. How many weighings are necessary in order to isolate the counterfeit coin? How much information about the position of the coin does every weighing furnish in this connection?

210. It is known that in an experiment with three outcomes having probabilities p, q, r, the entropy H is $\leqslant 1$. Prove that then $\max(p, q, r)\geqslant\frac{1}{2}$.

211. Let $\{p_k)$ be an arbitrary distribution, where

$$\sum_{k=1}^{\infty} kp_k = \lambda > 1.$$

Prove that

$$H = -\sum_{k=1}^{\infty} p_k \log p_k$$

is maximal when $p_k=(1/\lambda)(1-1/\lambda)^{k-1}$.

212. *Puzzle problem.* Estimate the indeterminacy inherent in the following weather forecast: either rain, or snow, will occur, or not occur. It is known that at a given time of the year 75% of all days are with precipitation, where rain and snow cannot fall simultaneously. Assume also that on a day with precipitation, snow and rain are equally probable.

213. The probability of the occurrence of event A with one trial equals p. The trials are repeated until the first occurrence of event A. Find the entropy of the number of trials and explain the nature of the change in entropy with the change of p.

214. Define the entropy of a random variable, which is subject to the binomial distribution law:

a) in the general case;
b) for $p=\frac{1}{2}$ and $n=10$.

215. Let $\xi_1, \xi_2, ..., \xi_n$ be a sequence of mutually independent random variables which take on the values zero and unity with probabilities p and $q=1-p$ respectively. Denote by $p(\mathbf{x})$, where $\mathbf{x}=(x_1,..., x_n)$, $x_i=0, 1$, the probability that $\xi_i=x_i$ for all i. Prove that

$$\frac{1}{n}\sum_{\mathbf{x}} p(\mathbf{x})\log p(\mathbf{x}) = -H(p) = p\log p + q\log q,$$

where the summation is over all possible vectors $\mathbf{x}$.

216. Prove that $H(\xi)$ does not surpass $\log n$, where n is the number of values taken on by ξ and that the maximum is attained for

$$p_k = \frac{1}{n} \quad (k = 1, 2, ..., n).$$

217.* Prove that
a) $H(\xi, \eta) \leqslant H(\xi) + H(\eta)$;
b) $I_\xi(\eta) = I(\xi) \geqslant 0$
and equality is attained if, and only if, ξ and η are independent.
Hint. Use the following inequality. If $\{p_i\}$ and $\{q_i\}$ $(i=1, 2, ..., n)$ are two systems of nonnegative integers, with $\sum p_i = \sum q_i = 1$, then

$$\prod_{i=1}^{n} \left(\frac{p_i}{q_i}\right)^{q_i} \leqslant 1,$$

where equality is attained if, and only if, $p_i = q_i$ for all i.

218. The probabilities that a signal is received or not received on the intake of a receiver are equal to α and $1-\alpha$ respectively. As a result of a hindrance, a signal received on the intake of a receiver can be apprehended on outgo with probability β and not be apprehended with probability $1-\beta$. In the absence of the signal on the intake it can, because of a hindrance, be apprehended on outgo with probability γ and not be apprehended with probability $1-\gamma$. Determine the quantity of information on the receipt or non-receipt of the signal on the intake by observation of the presence or absence of the signal on the outgo.

219. A signal X which is either x_1 or x_2 is transmitted along each of two duplicating channels. The *a priori* probabilities of x_1, x_2 are respectively p, $1-p$. Noise acts independently on the two channels, distorting transmission. We use y_1 to signify that signal x_1 was received along the first channel; the transmitted signal may have been x_1 or x_2. Similarly, y_2 signifies that x_2 was received along the first channel. We use z_1 and z_2 analogously in relation to the second channel. The matrices of conditional probabilities $a_{ik}=P\{y_k \mid x_i\}$ for the first channel and $P\{z_k \mid x_i\}$ for the second channel are the same and equal

$$\begin{pmatrix} a_{11} & a_{12} \\ a_{21} & a_{22} \end{pmatrix} = \begin{pmatrix} \varDelta & 1-\varDelta \\ 1-\delta & \delta \end{pmatrix}.$$

Find $I_{YZ}(X)$. Compute $I_{YZ}(X)$ for $p=0.5$ and $\varDelta=\delta=0.9$.

220. Under the conditions of the preceding problem, find the quantity of acquired information if in decoding the transmission one of the following rules is used:

a) If the signal x_1 was received on both channels, then it is assumed that the signal x_1 has been transmitted; in the remaining cases, it is assumed that the signal x_2 has been transmitted.

b) If the same signal has been transmitted on both channels, then it is assumed that this signal was actually transmitted. But if different signals were received, then it is assumed that the signal x_1 was transmitted.

221.* Prove that the following conditions define the function $H(p_1,\ldots,p_n)$ to within a constant factor, whose value serves only to determine the unit of the quantity of information:

a) $H(p, 1-p)$ is a continuous function of p on the segment $0 \leqslant p \leqslant 1$;

b) $H(p_1,\ldots,p_n)$ is a symmetric function in all of its variables;

c) if $p_n=q_1+q_2>0$, then

$$H(p_1,\ldots,p_{n-1},q_1,q_2) = H(p_1,\ldots,p_n) + p_n H\left(\frac{q_1}{p_n}, \frac{q_2}{p_n}\right);$$

d) $F(n)=H(1/n,\ldots,1/n)$ is a monotonically increasing function of n.

Remark. Condition d) is not necessary and is introduced only to facilitate the proof.

Hint. Prove successively the following assertions:

1) $H(1, 0) = 0$;

2) $H(p_1,\ldots,p_n, 0) = H(p_1,\ldots,p_n)$;

3) $H(p_1, ..., p_{n-1}, q_1, ..., q_m) = H(p_1, ..., p_n) + p_n H\left(\dfrac{q_1}{p_n}, ..., \dfrac{q_m}{p_n}\right)$,

where $p_n = q_1 + \cdots + q_m > 0$;

4) $H(q_{11}, ..., q_{1m_1}; ...; q_{n1}, ..., q_{nm_n}) =$

$$= H(p_1, ..., p_n) + \sum_{i=1}^{n} p_i H\left(\dfrac{q_{i1}}{p_i}, ... \dfrac{q_{im_i}}{p_i}\right),$$

where $p_i = q_{i1} + \cdots + q_{im_i}$;

5) $F(mn) = F(m) + F(n)$;

6) $F(n) = k \log n$, where k is an arbitrary constant;

7) For $p = r/s$,

where r, s are integers $H(p, 1-p) = k(p \log p + (1-p) \log(1-p))$.

See page 131 for the answers on problems 149–221.

4   *Basic limit theorems*

The problems of this chapter correspond to the material of §§12–15, 31–32, 41–43 of the textbook by B. V. GNEDENKO.

Normal and Poisson approximations to the binomial distributions.
Let $S_n = \sum_{i=1}^n \xi_i$, where the ξ_i are independent, identically distributed random variables, which take on the values 0 and 1 respectively with the probabilities p and $q = 1 - p$. In 2.2 the formula $P\{S_n = m\} = C_n^m p^m q^{n-m}$ ($0 \leqslant m \leqslant n$) was used extensively. It is not difficult to show that as $p \to 0$, $n \to \infty$, $np \to \lambda < \infty$,

$$P\{S_n = m\} \to \frac{\lambda^m}{m!} e^{-\lambda} = \pi(m).$$

A nonnegative integral random variable which takes on the value m with probability $\pi(m)$ is said to have a Poisson distribution with parameter λ.

The formula $P\{S_n = m\} \approx \pi(m)$ is called the *Poisson approximation to the binomial distribution*. It is usually used in the case when $p \leqslant 0.1$ and $npq \leqslant 9.$[1] In the case when $npq \geqslant 9$, the normal approximation, based on the de Moivre-Laplace theorem, which asserts that as $npq \to \infty$,

$$P\left\{a < \frac{S_n - np}{\sqrt{npq}} < b\right\} \to \frac{1}{\sqrt{2\pi}} \int_a^b \exp\left\{-\frac{u^2}{2}\right\} du$$

holds.

Estimates of the precision of these approximate formulas are contained in Problems 245, 246, 324.

Law of Large Numbers (LLN). We say that the LLN is applicable to the

---

[1] See A. HALD. *Mathematical statistics.*

sequence of random variables $\xi_1, \xi_2, ..., \xi_n, ...$ if, for an arbitrary $\varepsilon > 0$,

$$\lim_{n \to \infty} P \left\{ \left| \frac{1}{n} \sum_{i=1}^{n} \xi_i - \frac{1}{n} \sum_{i=1}^{n} M[\xi_i] \right| < \varepsilon \right\} = 1.$$

The majority of problems on the LLN are easily solved with the aid of *Chebyshev's inequality*. Let us recall it. For an arbitrary random variable ξ having dispersion, and for an arbitrary $\varepsilon > 0$,

$$P\{|\xi - M[\xi]| \geqslant \varepsilon\} \leqslant \frac{D[\xi]}{\varepsilon^2}.$$

The *Central Limit Theorem* (*CLT*) is a generalization of the de Moivre-Laplace theorem. If the sequence of mutually independent random variables $\xi_1, \xi_2, ..., \xi_n, ...$, for an arbitrary constant $\tau > 0$, satisfies the Lindeberg condition

$$\lim_{n \to \infty} \frac{1}{B_n^2} \sum_{k=1}^{n} \int_{|x - a_k| > \tau B_n} (x - a_k)^2 \, dF_k(x) = 0,$$

where

$$a_k = m[\xi_k]$$

and where

$$B_n^2 = \sum_{k=1}^{n} D[\xi_k],$$

then, as $n \to \infty$,

$$P\left\{ \frac{1}{B_n} \sum_{k=1}^{n} (\xi_k - M[\xi_k]) < x \right\} \to \frac{1}{\sqrt{2\pi}} \int_{-\infty}^{x} \exp\left\{ -\frac{u^2}{2} \right\} du$$

uniformly with respect to x. It is important to note that if all the ξ_i are identically distributed and $D[\xi] < \infty$, then the Lindeberg condition is satisfied.

Numerous problems on the proof of the applicability of the LLN and CLT are easily solved with the aid of characteristic functions (see the problems of Chapter 5).

The reader may wish to consult [2] and [13] for further discussion of these topics.

4.1 The de Moivre-Laplace and Poisson theorems

222. For an experimental verification of the law of large numbers the following experiments were performed at various times:

i) A coin was tossed 4040 times, heads came up 2048 times (Buffon).

ii) When a coin was tossed 12,000 times, the relative frequency of heads turned out to be 0.5016; in another experiment in tossing a coin 24,000 times the relative frequency of heads was 0.5005 (Pearson).

iii) Four coins were tossed 20,160 times and the combinations: four heads, three heads and a tail, two heads and two tails, one head and three tails, four tails came up the following number of times respectively: 1181, 4909, 7583, 5085, 1402 (V. I. Romanovsky). For each of the experiments, find the following:

a) the probability that the relative frequency of heads in an identical experiment with a fair coin will differ from $\frac{1}{2}$ by at least the amount observed;

b) assuming that an event the probability of whose occurrence equals 0.9999 is practically certain, find the practical upper bound of the possible deviation of the relative frequency of heads from the true probability of heads in each of the experiments.

223. It is known that the probability of the birth of a boy is approximately equal to 0.515. What is the probability that from among 10 thousand newly-born there will be no more boys than girls?

224. 200 persons attend a lecture on the theory of probability. Find the probability that k persons of those attending were born on May 1 and that l persons were born on November 7. Assume that the probability of birth on a fixed day equals $\frac{1}{365}$. Solve the problem for $k=1$ and $l=2$. Find the probability that the number of those born on either May 1 or November 7 is not greater than 2.

225. 1,359,671 boys and 1,285,086 girls were born in Switzerland from 1871 to 1900. Are these data compatible with the assumption that the probability of the birth of a boy equals 0.5? 0.515?

Hint. Let X be a random variable with the normal distribution of mean 0 and dispersion 1. Let p = probability of the birth of a boy, $n = 2,644,757$ and $m = 1,359,671$. Say the data are compatible if

$$P\left[X > \frac{\dfrac{m}{n} - p}{\sqrt{\dfrac{pq}{n}}} \right] \geqslant .001$$

and incompatible otherwise. Evaluate this probability for $p = .5$ and $p = .515$.

226. [See Problem 118]. With the purpose of determining π experimentally, a needle was thrown 5000 times and intersected lines 2532 times (Wolf and Tsyurikhe), where $2a = 45$ mm and $2l = 36$ mm. With what error was π determined? How many throwings of the needle are necessary in order that, with $a = l$, the probability that π will be calculated with an error not exceeding 0.001, to be 0.5; 0.95; 0.999?

227. Among seeds of wheat, 0.6% are weed seeds. What is the probability that, in a random selection of 1000 seeds, one finds not less than 3 weed seeds? Not more than 16 weed seeds? Exactly 6 weed seeds?

228. The per cent content of cementite on a metallographical edge was determined with the aid of a spike which touched the edge in a random way and the number of fallings of the spike on the structure being studied was observed. What should have been the per cent content of cementite in order that, with probability greater than 0.95, in 400 observations the spike fell on the cementite more than 100 times?

229. A book of 500 pages contains 50 errors. Estimate the probability that there are not less than 3 errors on a randomly selected page.

230. For persons having survived to their twentieth year, the probability of death on the twenty-fifth year of life equals 0.006. A group of 10,000 persons aged 20 years is insured, and every person insured contributes 1.2 rubles in premium a year. In the case of the death of an insured person, the insurance company pays out to the heirs 100 rubles. What is the probability that a) at the end of the year the insurance company will be in the red; b) its income will exceed 6000 rubles; 4000 rubles?

231. Many botanists performed experiments on the crossing of yellow (hybrid) peas. According to a known hypothesis of Mendel, the probability of the appearance of a yellow pea in such experiments equals $\frac{1}{4}$. In 34,153 crossing experiments, a yellow pea was obtained in 8506 cases.

a) Assuming that the probability of obtaining a green pea in all experiments was constant and equal to $\frac{3}{4}$, find the probability of the inequality

$$0.245 < v < 0.255,$$

where v is the frequency of the appearance of a green pea.

b) Assuming that the probability of obtaining a green pea in all experiments is equal to $\frac{3}{4}$, find the probability that upon performing 34,153 similar experiments, the deviation of the relative frequency from $\frac{3}{4}$ will be greater in absolute value than that obtained initially.

c) Assuming that the probability of obtaining a green pea in all experiments was equal to $\frac{3}{4}$, find how many analogous experiments it is necessary to make in order that with probability 0.99 one could assert that the deviation of the relative frequency from $\frac{3}{4}$ will not exceed 0.01.

232. In investigating the influence of radiation on the division of yeast cells preparations were studied under a microscope after radiation and incubation in an incubator. On each of 3 object slides there were counted 400 microcolonies and those which contained 2–4 cells (inactivation after 1–2 gemmations) were noted. On the first slide, there were found 220 such microcolonies, on the second 190, and on the third 210. Are these data compatible with the assumption that, with the applied dosage, each cell will, with probability $\frac{1}{2}$, be inactivated after 1–2 gemmations? What would be the probability under this same assumption of obtaining in the investigation of 10 slides at least one with more than 230 inactivations after 1–2 gemmations?

233. In performing a telepathic experiment, the inductor, independently of the preceding trials, chooses with probability $\frac{1}{2}$ one of 2 objects and thinks of it, and the recipient guesses what object the inductor is thinking about. Such an experiment was performed 100 times, and 60 correct answers were obtained. What is the probability of coincidence in one experiment, under the assumption that there is no telepathic connection between the inductor and recipient? Can one ascribe to the result obtained a purely random coincidence or not?

234. It is known that the probability of issuing a drill of high brittleness (a reject) equals 0.02. Drills are packed in boxes of 100 each. What is the probability that:

a) there are no defective drills in a box?

b) the number of defective drills turns out to be no greater than 2? What is the smallest quantity of drills that need to be put in a box in order that, with probability not less than 0.9, there are in it not less than 100 good drills in the box?

235. *Puzzle problem.* How many raisins on the average must caloric buns contain in order that the probability of having at least one raisin in a bun be not less than 0.99?

236. A Geiger-Müller counter and a source of radioactive particles are so situated that the probability that a particle emanating from the radioactive source is registered by the counter equals 1/10,000. Assume that during the time of observation, 30,000 particles emanated from the source. What is the probability that the counter:

a) registered more than 10 particles?

b) did not register a single particle?

c) registered exactly 3 particles?

237. Under the conditions of the preceding problem, what is the smallest number of particles that must emanate from the source in order that, with probability greater than 0.99, the counter registers more than 3 particles?

238. Assume that in the composition of a book there exists a constant probability $p = 0.0001$ that an arbitrary letter will be set incorrectly. After the composition, the proofs are read by a proofreader who discovers any particular error with a probability of $q = 0.9$. After the proofreader, the author discovers the remaining errors with probability $r = 0.5$ for each. Find the probability that in a book with 100 thousand printing symbols there remain after this not more than 10 unnoticed errors.

239. Using the de Moivre-Laplace theorem, prove the Bernoulli theorem which asserts that for an arbitrary $\varepsilon > 0$, the probability that the deviation of the relative frequency of success from the probability of success will be greater than ε, tends to zero when the number of trials tends to infinity.

240. A theater, accommodating 1000 persons, has two different entrances.

There is a cloakroom at each of the entrances. How many places should there be in each of the cloakrooms in order that, on the average in 99 cases of 100, all the spectators can leave their coats at the cloakroom of that entrance through which they entered? It is assumed that the spectators arrive in pairs and each pair independently of the others is equally likely to choose either entrance. By how many can one decrease the number of places in the cloakroom if the spectators arrive singly and, independently of one another, with equal probability choose either of the entrances?

241. A certain machine consists of ten thousand parts. Each part, independently of the other parts, can turn out to be in disrepair with probability p_i, where for $n_i = 1000$ parts, $p_1 = 0.0003$, for $n_2 = 2000$ parts, $p_2 = 0.0002$, and for $n_3 = 7000$ parts, $p_3 = 0.0001$. A machine does not work if at least two of its parts are in disrepair. Find the probability that the machine will not work.

242. To check the effect of a new medicine on blood pressure, the pressure of 100 patients was measured before and after administration of the medicine. In this connection, it turned out that in 32 cases the pressure increased after administration of the medicine and in 68 cases it decreased. Can one assume that it has been established that this medicine influences blood pressure? What is the probability that purely random variations of the pressure cause at least as large a deviation from 50?

243. In certain countries of Western Europe, from the seventeenth century up to the imperialist war of 1914, there existed the following government lottery: the lottery contained 90 numbers of which in each successive drawing there came out any 5 numbers; the player had the right to make in advance a bet on an arbitrary number or on a group of numbers; if all the numbers he wrote down turned out to be among the five which were drawn, then, in exchange for the bet, he obtained a prize; the prize exceeded the bet by 15 times if he wrote down one number; by 270 times if he wrote down two numbers; by 5,500 times if he wrote down three numbers; by 75,000 times if he wrote down four numbers; by 1,000,000 times if he wrote down five numbers.

Find the mean value of the winnings of a player who writes down one number, two numbers, ..., five numbers.

Assume that 100,000 persons made a wager on three numbers. Find the probability that the number of winners among them exceeds 10.

244. Using the Stirling formula, show that as $\lambda \to \infty$, for an arbitrary fixed n,

$$n\sqrt{\lambda} \cdot \left| \frac{\lambda^n}{n!} e^{-\lambda} - \frac{1}{\sqrt{2\pi\lambda}} \exp\left\{ -\frac{1}{2\lambda}(n-\lambda)^2 \right\} \right| \to 0.$$

245. Using Stirling's formula, show that if $np(1-p) \geqslant 25$, then

$$\sum C_n^k p^k (1-p)^{n-k} = \frac{1}{\sqrt{2}} \int_a^b \exp\left\{ -\frac{x^2}{2} \right\} dx + \frac{1-2p}{6\sqrt{2\pi np(1-p)}} \times$$

$$\times \left[(1-b^2)\exp\left\{ -\frac{b^2}{2} \right\} - (1-a^2)\exp\left\{ -\frac{a^2}{2} \right\} \right] + R,$$

where

$$|R| < \frac{0.13 + 0.18\,|1-2p|}{np(1-p)} + \exp\left\{ -\tfrac{3}{2}\sqrt{np(1-p)} \right\}$$

and the summation is over k lying within the bounds

$$np + \tfrac{1}{2} + a\sqrt{np(1-p)} \leqslant k \leqslant np - \tfrac{1}{2} + b\sqrt{np(1-p)}.$$

246. Let

$$W_k = C_n^k p^k q^{n-k}, \qquad V_k = \frac{\lambda^k e^{-\lambda}}{k!}, \qquad \Delta_k = \frac{|W_k - V_k|}{W_k}.$$

Prove that for an arbitrary $\lambda = np$, the conditions $k^2 \leqslant n\varepsilon$ and $\lambda^2 \leqslant n\varepsilon$, where $\varepsilon < \tfrac{1}{9}$, imply that $\Delta_k \leqslant 1.2\varepsilon$.

4.2 Law of Large Numbers and convergence in probability

247. Prove that if $\eta_n \overset{p}{\to} a$[1]) and $0 < \zeta_n \overset{p}{\to} b > 0$ where $|\eta_n/\zeta_n| < C$, then $M[\eta/_n\zeta_n] \to a/b$.

248. Prove that if the function f is continuous at the point a and if the sequence of random variables $\xi_n \overset{p}{\to} a$, then $f(\xi_n) \overset{p}{\to} f(a)$.

---

[1]) The expression $\xi_n \overset{p}{\to} a$ means that the sequence ξ_n converges to a in probability i.e., that for an arbitrary $\varepsilon > 0$, $P\{|\xi_n - a| > \varepsilon\} \to 0$ $(n \to \infty)$.

249. Prove that if $|\xi_n| \leqslant k$ and $\xi_n \overset{P}{\to} a$, then also $M[\xi_n] \to a$. Show that the requirement $|\xi_n| < k$ is not essential. How can it be weakened?

250. Let $F(x) = P\{\xi_n < x\}$ be continuous and let the random variable $\eta_n \overset{P}{\to} 1$ in probability. Prove that:

a) $\lim\limits_{n \to \infty} P\{\xi_n + \eta_n < x\} = F(x-1)$;

b) $\lim\limits_{n \to \infty} P\left\{\dfrac{\xi_n}{\eta_n} < x\right\} = F(x)$.

251. Let f be a continuous function which is monotonically increasing on the segment $[0, \infty]$, with $f(0) = 0$ and $\sup_{0 \leqslant x} f(x) < \infty$.

Prove that the condition $\lim_{n \to \infty} M[f(|\xi_n|)] = 0$ is necessary and sufficient in order that $\xi_n \overset{P}{\to} 0 \, (n \to \infty)$ in probability.

252. Prove the LLN using the Chebyshev inequality.

253. Prove that for every continuous function $f(x)$ on the segment $[0, 1]$ the Bernstein polynomials

$$B_n(x) = \sum_{k=0}^{n} f\left(\frac{k}{n}\right) C_n^k x^k (1-x)^{n-k}$$

tend uniformly in x to $f(x)$ as $n \to \infty$. (Weierstrass' theorem)

254. Let $f(m) \ (m = 1, 2, \ldots)$ be an arbitrary sequence of real numbers; $v_n\{\ldots\}$ is the frequency of all natural numbers $m \leqslant n$ subject to the conditions which are written in the braces,

$$M_n = \frac{1}{n} \sum_{m=1}^{n} f(m); \quad D_n = \frac{1}{n} \sum_{m=1}^{n} (f(m) - M_n)^2.$$

Let $\Psi(n)$ be an arbitrary function which increases indefinitely as $n \to \infty$. Prove the following analogue of the law of large numbers:

$$v_n\{|f(m) - M_n| \leqslant \Psi(n) \sqrt{D_n}\} \to 1 \quad (n \to \infty).$$

255. Let $S_n = \xi_1 + \xi_2 + \cdots + \xi_n$. Prove that if for all n, $|S_n| < Cn$ and $D(S_n) > \alpha n^2$, then the law of large numbers is not applicable to $\{\xi_k\}$.

256. Let $\{\xi_k\}$ be a sequence of random variables such that ξ_k can depend only on ξ_{k-1} and ξ_{k+1} but does not depend on all the other ξ_j. Show that the law of large numbers is satisfied if $D[\xi_k] < C < \infty$.

257. If the joint distribution of the quantities $\xi_1, \xi_2, ..., \xi_n$ is defined for every n, where $D[\xi_j] < C < \infty$, and the covariance is negative, then the law of large numbers is applicable. Prove that if the condition $r_{ik} = \text{cov}(\xi_i, \xi_k) \leqslant 0 \ (k \neq i)$ is replaced by the assumption that $r_{ik} \to 0$ uniformly as $|i-k| \to \infty$ then the law of large numbers is also applicable.

258. Let $\Xi_i \ (i = 1, 2, ...)$ be independent identically distributed random variables with $M[\xi] = a > 0$ and $D[\xi] = \sigma^2$. As follows from the central limit theorem, in this case

$$\frac{\left(\sum_{i=1}^{n} \xi_i - na\right)}{\sigma \sqrt{n}}$$

converges weakly[1]) to the normal law with parameters $[0, 1]$. Using the law of large numbers, prove that

$$2 \frac{\sqrt{a}}{\sigma} \left(\sqrt{\left|\sum_{i=1}^{n} \xi_i\right|} - \sqrt{na}\right)$$

also converges to the normal law with parameters $[0, 1]$ as $n \to \infty$.

259. Let $\xi_1, \xi_2, ...$ be independent identically distributed random variables which take on the values 0 and 1 with probabilities p and $1-p$ respectively. We denote by $\mathcal{M}_n$ the collection of all possible sequences of zeros and ones of length n. Prove that, for arbitrary $\varepsilon > 0$ and $\delta > 0$, there exists an n_0 such that for an arbitrary $n > n_0$, $\mathcal{M}_n$ decomposes into two classes $\mathcal{M}_n^1$ and $\mathcal{M}_n^2$ such that

1) $P\{(\xi_1, \xi_2, ..., \xi_n) \in \mathcal{M}_n^1\} < \varepsilon$;

2) $p(\mathbf{x}) = P\{\xi_1, ..., \xi_n = x_n\}$, the probabilities of the sequences $\xi = \{\xi_1, ..., \xi_n\} \in \mathcal{M}_n^2$, satisfy the inequality

$$\left| -\frac{1}{n} \log p(\mathbf{x}) - H \right| < \delta, \quad \text{where} \quad H = -p \log p - q \log q.$$

260. The notations are the same as in the preceding problem. Arrange

----

the sequences $\{x\}$ belonging to $\mathcal{M}_n$ in order of decreasing probabilities $p(x)$. Denote by $m(r)$ the number of sequences which we must take from $\mathcal{M}_n$, starting with the most probable sequence, in order to accumulate the total probability r for the sequences we have taken. Prove that, for $0 < r < 1$,

$$\lim_{n \to \infty} \frac{\log m(r)}{n} = H.$$

4.3 Central Limit Theorem

261. A die is thrown 1000 times. Find the limits within which the number of eyes coming up will lie with probability greater than 0.99.

262. In firing at a target, a marksman scores at each shot either 10, 9, 8, 7 or 6 with respective probabilities 0.5, 0.3, 0.1, 0.05, 0.05. He fires 100 shots. What is the probability that his aggregate score exceeds 980? 950?

263. In setting up a statistical counting, it was necessary to add 10^4 numbers, each of which was rounded off with accuracy up to the 10^{-m} degree. Assuming that the errors arising from rounding off the numbers are mutually independent and uniformly distributed on $(-0.5 \cdot 10^{-m}, 0.5 \cdot 10^{-m})$, find the limits in which the total error will lie with probability greater than 0.997.

264. On the segment $[0, 1]$, a number ξ is chosen at random and expanded in a decimal fraction $\xi = \sum_{n=1}^{\infty} e_n(\xi)/10^n$. Prove that the distribution $S_n = \sum_{k=1}^{n} e_k(\xi)$, for a suitable normalization, tends to the normal law as $n \to \infty$.

265. *From the history of measure.* The measure of length "foot", as is clear from the nomenclature, has a direct relation to the human foot. This is the length of a step. But, as is known, the dimensions of feet are various. In the XVI century, the Germans came out of this situation in the following way. On a holiday, they lined up the sixteen men who first came out of the church. The sum of the lengths of their left feet was divided by sixteen. The mean length was then the "correct and legal foot". It is known that the dimension of the foot of a mature man is a random variable, having a normal distribution with mean value 262.5 mm and standard deviation $\sigma = 12$ mm. Find the probability that two "correct and legal" values of the foot, defined with respect to two different groups of men, differ by more than 5 mm. How many men should have been taken in order that with probability greater than 0.99 the mean dimension of their feet differ from 262.5 mm by less than 0.5 mm?

266. Let ξ_i denote the time between two successive mutual collisions, the i-th and the $(i+1)$-st, of the molecule described in Problem 141. Find the limiting distribution as $n \to \infty$ for $\sum_{i=1}^{n} \xi_i$ (see also Problem 154).

267. Establish whether the LLNs and the CLT will be satisfied for independent random variables ξ_k with distributions defined in the following way $(k \geqslant 1)$:

a) $P(\xi_k = \pm 2^k) = \frac{1}{2}$

b) $P(\xi_k = \pm 2^k) = 2^{-(2k+1)}$; $\quad P(\xi_k = 0) = 1 - 2^{-2k}$;

c) $P(\xi_k = \pm k) = \frac{1}{2} k^{-1/2}$; $\quad P(\xi_k = 0) = 1 - k^{-1/2}$.

268. Find the distribution of $\xi_n = \sum_{i=1}^{n} \xi_i$, where the ξ_i are independent, identically distributed random variables, each of which has a Poisson distribution with parameter 1. Find the limit distribution for $\xi_n - n/\sqrt{n}$ as $n \to \infty$.

269. Prove that if f is a function which is continuous and bounded on $[0, \infty]$, then for $h > 0$,

$$\lim_{n \to \infty} \sum_{k=0}^{\infty} f\left(x + \frac{k}{n}\right) \frac{(nh)^k}{k!} e^{-hn} = f(x + h)$$

(see the preceding problem).

270. Let ξ denote the resistance to breaking of some metallic column. Assume that all columns participating in some production process have the same resistance to breaking $\xi_0 = a_0$. During the first "stage" of the process all columns are subjected, for example, to a thermal treatment, the purpose of which is to modify the resistance to breaking of each column from a_0 to a_1. In view of the random variations in treatment, each of the columns does not acquire the exact desired resistance, but the results ξ_1 oscillate at random about a_1. After this, the columns are subjected to a second stage of treatment, which is intended to modify

ξ_1 from a_1 to a_2, in which connection the result ξ_2 will deviate randomly from a_2, and so on. Below, there are given two simple mathematical models of such processes.

a. Assume that the successive deviations $\Delta_k = \xi_k - \xi_{k-1}$ do not have a systematic error (i.e., $M[\Delta_k] = a_k - a_{k-1}$) and are mutually independent with $|\Delta_k - (a_k - a_{k-1})| \leqslant b$. For a sufficiently large n find an $l_n = o(\sqrt{n})$ such that, independently of the distribution Δ_k ($k = 1, 2, ..., n$),

$$P\{|\xi_n - a_n| < l_n\} \geqslant 0.95.$$

b. Now assume that the relative deviations at each stage of the treatment, $\tau_n = \xi_n / \xi_{n-1} - 1$ are independent, do not have a systematic error and are identically distributed, whereby $|\tau_n| < \varepsilon$.

Find the limit distribution for the corresponding normalization for $\ln \xi_n$ as $n \to \infty$. Estimate for $n = 100$ and $\varepsilon = \frac{1}{200}$ the probability that $0.905 < n/a_0 < 1.105$.

271. A die is thrown for as long as the total sum of eyes coming up does not exceed 700. Estimate the probability that for this more than 210 tosses are required; less than 180 tosses; from 190 to 210 tosses.

272. A controller checks one after another the parts of some production. At the first step of the check, which lasts 10 secs., he either evaluates the part at once, or he makes the decision that the check must be repeated. The repeated check lasts 10 secs.; as a result of it, it is obligatory to make a decision about the quality of the production. Find the probability that during a 7-hour work day the controller checks more than 1800 parts; more than 1600 parts; not less than 1500 parts. It is assumed that every part, independently of the other parts, is subjected to a repeated check with a probability of 0.5.

273. Let V be a region in an s-dimensional space, having unit volume, and let $|f(\mathbf{x})| < a$ be a function defined everywhere in the region V. In order to calculate $I = \iiint_V f(\mathbf{x}) \, dV$ by the Monte-Carlo method, we proceed in the following way: N points $\mathbf{x}_1, ..., \mathbf{x}_N$ are located at random, independently one after the other, in the region V and for an approximate estimate of the integral we take

$$I_n = \frac{1}{N} \sum_{i=1}^{N} f(\mathbf{x}_i).$$

What is $M[I_N]$ equal to? Estimate $D[I_n]$. Find the limiting distribution for $\sqrt{N(I_N - I)}$ as $N \to \infty$.

274. The independent quantities $\xi_1, \xi_2, \ldots$ have the same distribution with $M[\xi_i] = 0$ and $D[\xi_i] = 1$. Show that the quantities

$$\eta = \sqrt{n}\,\frac{\xi_1 + \cdots + \xi_n}{\xi_1^2 + \cdots + \xi_n^2} \quad \text{and} \quad \zeta = \frac{\xi_1 + \cdots + \xi_n}{\sqrt{\xi_1^2 + \cdots + \xi_n^2}}$$

are each asymptotically normal with parameters $(0, 1)$.

275. Let $\xi_i (i = 1, 2, \ldots, n+1)$ be independent with each normal distributed with mean 0 and dispersion 1. Set

$$\chi_n^2 = \sum_{i=1}^{n} \xi_i^2 \quad \text{and} \quad \tau_n = \frac{\xi_{n+1}}{\frac{1}{n}\chi_n^2}$$

Find the limiting distribution of χ_n^2 and τ_n as $n \to \infty$.

276. *Puzzle problem.* On each autobus ticket there is a six-place number. The ticket is called "lucky" if the sum of the first three digits of its number coincides with the sum of the last three digits. Assuming that all numbers of the tickets from 000000 to 999999 are equally probable, find the probability that a randomly chosen ticket turns out to be "lucky". Find this probability directly and also with the aid of the local limit theorem[1]. Compare the results obtained.

277.* Let $S_n = \xi_1 + \xi_2 + \cdots + \xi_{\mu(n)}$ be the sum of a random number $\mu(n)$ of random variables ξ_i, where ξ_i and $\mu(n)$ are independent. $|\xi_i| < c$, $M[\xi_i] = a$, $D[\xi_i] = \sigma^2$; $\mu(n)$ is an integral nonnegative random variable with $M[\mu(n)] = n$ and $D[\mu(n)] \leqslant n^{1-\varepsilon}$, where $\varepsilon > 0$. Find the limiting distribution of

$$\frac{S_n - na}{\sigma \sqrt{n}} \quad \text{as} \quad n \to \infty.$$

278. A series of a very large number n of trials is considered. The trials are decomposed into groups of 3. The first two trials of each group are independent, the probability of the occurrence of the event A at either

---

[1]) See GNEDENKO. *Theory of probability*, §12, pp. 94–103, Chelsea (1967).

being $\frac{1}{2}$. The outcome of the third trial in a group is predicted from the results of the preceding two trials by the restriction that the number of occurrences of A in each group must be even (0 or 2). Show that the number m of occurrences of the event A in the entire series n is subject to the same Laplace-Gauss limit law as if all the trials (as in the Bernoulli scheme) were independent.

279. A person stands on the street and sells newspapers. Assume that each of the people passing by buys a newspaper with probability $\frac{1}{3}$. Let ξ denote the number of people passing past the seller during the time until he sells his first 100 copies of the newspaper. Find the distribution of ξ.

280. From the numbers $1, 2, ..., N$, n are chosen at random. Denote them by $m_1, ..., m_n$. Set $\xi_i = 0$ if $m_i = 0$ modulo 3, $\xi_i = 1$ if $m_i = 1$ modulo 3, and $\xi_i = -1$ if $m_i = 2$ modulo 3. Find the probability that $S = \sum_{i=1}^{n} \xi_i = k$.

Prove that when n and N tend to infinity so that $n = o(N)$, $(1/\sqrt{n}) S_n$ converges in probability to the normal distribution with parameters $(0, \sqrt{\frac{2}{3}})$.

See page 134 for the answers on problems 223–279.

5 *Characteristic and generating functions*

The material of this chapter corresponds basically to Chapter 7 of the textbook by B. V. GNEDENKO. To solve the problems of the first two sections, it is sufficient to know the definitions of a characteristic function and of a generating function.

The *characteristic function* (c.f.) of a random variable ξ is defined by $f(t) = M[\exp\{it\xi\}]$. The *generating function* (g.f.) of a sequence of numbers $\{a_r\}$ $(r = 0, 1, 2, \ldots)$ is defined with the aid of the formula

$$\phi(\mathfrak{z}) = \sum_{r=0}^{\infty} a_r \mathfrak{z}^r.$$

In the case when all the a_r are the probabilities that the random variable ξ equals r, $\phi(\mathfrak{z})$ is called the *generating function of* ξ. One of the basic properties of a c.f. and of a g.f. is that given it, one can uniquely reconstruct the distribution. In the third section, we systematically use also the theorem to the effect that a necessary and sufficient condition for a sequence of d.f.'s $F_n(x) \to F(x)$ as $n \to \infty$ at every point of continuity of the d.f. $F(x)$ is that for every fixed t the corresponding c.f.'s $f_n(t) \to f(t)$, as $n \to \infty$, where $f(t)$ is the c.f. of $F(x)$. One must pay special attention to Problems 282, 284, 289, 302, 303, 317–320. In a number of problems, the normal distribution with parameters $(0, 1)$ is denoted for brevity by $N(0, 1)$.

Convolution. If $F(x)$ and $G(x)$ are distribution functions, their convolution, $H(y)$, is defined by

$$H(y) = \int_{-\infty}^{\infty} F(y - x)\, dG(x).$$

$H(y)$ is a distribution function and $H(y)$ is denoted by $F(x) * G(x)$. For more detail see [9], [11], [12] and [13].

5.1 Calculation of c.f.'s and g.f.'s

281. Find the laws of distribution to which the following characteristic functions correspond:

$$\cos t; \quad \cos^2 t; \quad \sum_{k=0}^{\infty} a_k \cos kt, \quad \text{where} \quad a_k \geqslant 0 \quad \text{and} \quad \sum_{k=0}^{\infty} a_k = 1.$$

282. Calculate the c.f. for the following distribution laws:
a) uniform distribution in the interval $(-a, a)$;
b) binomial distribution;
c) Poisson distribution;
d) Cauchy distribution:

$$p(x) = \frac{1}{\pi} \cdot \frac{1}{1 + x^2};$$

e) exponential distributions with densities

$$p_1(x) = \begin{cases} 0 & (x < 0) \\ a\,e^{-ax} & (x > 0), a > 0, \end{cases}$$

and

$$p_2(x) = \tfrac{1}{2} e^{-|x|};$$

f) normal distribution:

$$p(x) = \frac{1}{\sqrt{2\pi}\sigma} e^{-(x-a)^2/2\sigma^2}.$$

283. Find the distribution laws to which the following generating functions correspond:

a) $\tfrac{1}{4}(1 + z^2)^2$; b) $\tfrac{1}{2}(1 - \tfrac{1}{2}z)^{-1}$;
c) $e^{\lambda(z-1)}$; d) $(\tfrac{1}{3} + \tfrac{2}{3}z)^n$.

284. Let ξ be a nonnegative integral variable with generating function $\phi(\mathfrak{z})$. Find the generating functions for variables:

$$\xi + 1 \quad \text{and} \quad 2\xi;$$

sequences:

$$P\{\xi \leqslant n\}; \quad P\{\xi < n\};$$
$$P\{\xi \geqslant n\}; \quad P\{\xi > n + 1\}; \quad P\{\xi = 2n\}.$$

285. Consider a Bernoulli sequence of trials. Let U_n be the probability that the first combination success-failure appears at the $(n-1)$-st and n-th trials. Find the generating function, mean value and dispersion of U_n, if the probability of success is p.

286. Let u_n be the probability that the number of successes in a sequence of n Bernoulli trials is divisible by 2. Prove the recursion formula $u_n = = qu_{n-1} + p(1 - u_{n-1})$.

Derive from this the generating function, and from it the exact formula for u_n.

287. A sequence of Bernoulli trials to the first failure inclusive will be called a cycle. Find the generating function and the distribution of the probability of the general number of successes in r cycles.

288. Let u_n be the probability that the number of successes in n Bernoulli trials is divisible by three. Find the recursion relation for u_n, and from it find the generating function.

5.2   Connection with properties of a distribution

289. Prove that a necessary and sufficient condition for a law of distribution of a random variable to be symmetric is that the characteristic function be real.

Remark. A distribution law is symmetric if

$$F(x) = 1 - F(-x + 0).$$

290. A random variable ξ is called lattice-type if one can represent the possible values of ξ in the form $a_r = a + k(r)h$, where $k(r)$ is an integer. The maximal value of h is called the *maximal mesh of the distribution*. Prove that if for some $t \neq 0, f(t)$ is the characteristic function of the random variable ξ with absolute value equal to 1, then ξ is a lattice-type random variable. Find the maximal mesh of the distribution if $f(t)$ is known.

291.* Suppose the random variable $|\xi| \leqslant A$ has density $p(x) < M$.

a) Prove that as $t \to \infty$, the characteristic function of ξ tends to zero.

b) Show that an absolute constant C can be found, not depending on A and M, such that

$$\max_{|t| > \pi/A} |f(t)| < 1 - \frac{C}{(AM)^2}.$$

292. Let ξ be a random variable with generating function $\phi(z)$ and suppose $\sum P\{\xi = n\} z^n$ converges for some $z_0 > 1$. Prove that in this case all moments $m_r = M[\xi^r]$ exist and the generating function $\psi(z)$ of the sequence $m_r/r!$ converges at least for $|z| < \log z_0$. Moreover,

$$\psi(z) = \sum_0^\infty \frac{m_r}{r!} z^r = \phi(e^z).$$

293.* Prove that if $F(x)$ is a distribution function and $f(t)$ is the corresponding characteristic function, then:

a) for an arbitrary x,

$$\lim_{T \to \infty} \frac{1}{2T} \int_{-T}^{T} f(t) e^{-itx} \, dt = F(x + 0) - F(x - 0);$$

b) $$\lim_{T \to \infty} \frac{1}{2T} \int_{-T}^{T} |f(t)|^2 \, dt = \sum_\nu (F\{x_\nu + 0\} - F\{x_\nu - 0\}),$$

where the x_ν are the abscissas of the jumps of the function $F(x)$.

294.* Prove that if $M[\xi] = 0$, then

$$M[|\xi|] = \frac{1}{\pi} \int_{-\infty}^{\infty} \frac{1 - Rf(t)}{t^2} \, dt,$$

where $f(t)$ is the c.f. of ξ and $R\phi$ is the real part of the function ϕ. If, furthermore, $D[\xi]$ exists, then

$$M[|\xi|] = -\frac{2}{\pi} \int_0^{\infty} \frac{R(f'(t))}{t} \, dt.$$

295. The distribution law $F(x)$ is called *stable* if, for arbitrary a_1, b_1; a_2, b_2 there exist a_3, b_3 such that

$$F(a_1 x + b_1) * F(a_2 x + b_2) = F(a_3 x + b_3).$$

Explain which of the following distribution laws belong to the stable type:

a) improper distribution law (distribution law with one point of growth);

b) binomial distribution law;

c) Gauss law;

d) Poisson law;

e) Cauchy law.

296. Prove that in adding independent random variables with zero means, the third central moments are summed, and the fourth are not.

297. Let ξ be a random variable, having a Poisson distribution with parameter v. If we consider the parameter v as a random variable with probability density

$$\frac{\alpha^\lambda}{\Gamma(\lambda)} x^{\lambda-1} e^{-\alpha x} \quad (x > 0), \quad \text{see page 19}$$

then the probability that ξ takes on a given value k equals

$$\int_0^\infty \frac{x^k}{k!} e^{-x} \frac{\alpha^\lambda}{\Gamma(\lambda)} x^{\lambda-1} e^{-\alpha x} \, dx =$$

$$= \left(\frac{\alpha}{1+\alpha}\right)^\lambda \frac{(-1)^k}{(1+\alpha)^k} \cdot \frac{(-\lambda)(-\lambda-1)\ldots(-\lambda-k+1)}{k!}.$$

Find the c.f., mean value and dispersion of this distribution, which is called the *negative-binomial distribution*.

298. A die is tossed n times. Let $\xi_1, \xi_2, \ldots, \xi_6$ denote respectively the number of occurrences of a one, a two, ..., a six. Using the multidimensional generating functions, find:

a) $\operatorname{cov}(\xi_1, \xi_2)$;

b) $M\left[(\xi_1 - M[\xi_1])^k (\xi_2 - M[\xi_2])^l\right]$, where k, l are integers $\geqslant 0$;

c) $M\left[\prod_{i=1}^{6}\left(\xi_i - \dfrac{n}{6}\right)\right]$;

d) $M[(1 + \xi_1)^{-1}]$;

e) $M[(\xi_1 - \xi_2)^3]$;

f) $M[[(\xi_1 - \xi_2)^2 + 1]^{-1}]$;

g) $M\left[\dfrac{\xi_1 + \xi_2}{\xi_2 + \xi_3 + 1}\right]$.

5.3 Use of the c.f. and g.f. to prove the limit theorems

299. Let $F(x)$ and $F_n(x)$ $(n=1, 2, ...)$ be nonnegative integral distribution functions, and let $\phi(z)$ and $\phi_n(z)$ $(n=1, 2, ...)$ be the generating functions corresponding to them. Prove that $\phi_n(z) \to \phi(z)$ $(n \to \infty)$ implies that $F_n(x) \to F(x)$ $(n \to \infty)$ uniformly with respect to x.

300. Use generating functions to show that, as $n \to \infty$ and $np \to \lambda < \infty$, the binomial distribution converges to the Poisson distribution.

301. Use c.f.s to show that, as $\lambda \to \infty$,

$$\sum_{\lambda + a\sqrt{\lambda} < k \leqslant \lambda + b\sqrt{\lambda}} \frac{\lambda^k}{k!} e^{-\lambda} \to \frac{1}{\sqrt{2\pi}} \int_a^b e^{-u^2/2} \, du .$$

302. Formulate in the language of c.f.'s a necessary and sufficient condition for a sequence of independent random variables $\xi_1, \xi_2, ..., \xi_n$ to be subject to the law of large numbers. The c.f.'s $f_i(t)$ of ξ_i are given.

303. Let $\xi_1, \xi_2, ...$ be a sequence of mutually independent and identically distributed random variables. Prove that: a necessary and sufficient condition for some constant C, that

$$\frac{1}{n} \sum_{i=1}^{n} \xi_i \xrightarrow{p} C$$

is that the c.f. of ξ_i be differentiable at the point $t = 0$.

304. *Moment generating function. A moment generating function* is the function of a real variable

$$m(t) = \int\limits_{-\infty}^{\infty} \exp\{tx\}\, dF(x).$$

Prove that: a) if two moment generating functions are equal, then the corresponding distribution functions are also equal;

b) if the sequence $m_n(t) = \int_{-\infty}^{\infty} \exp\{tx\}\, dF_n(x)$ converges to $m(t)$ for every value of t, then $F_n(x) \to F(x)$ at every point of continuity of $F(x)$.

305. Suppose $\xi_k \ (k=1, 2, \ldots)$ are independent and that for every k,

$$P\{\xi_k = k^{\alpha}\} = P\{\xi_k = -k^{\alpha}\} = \tfrac{1}{2}.$$

Use the result of Problem 302 to explain for what α the LLN is applicable to the sequence ξ_i. For what α is the CLT applicable?

306. Prove that if for a sequence of independent random variables $\{\xi_i\}$ there exist numbers $\alpha > 1$ and C such that $M[|\xi|^{\alpha}] \leqslant C$, then the LLN is applicable to the sequence (*Markov's theorem*).

307. $2n+1$ points are located at random on the segment $[0, 1]$, independently of one another. We denote by ξ^*_{n+1} the coordinate of the $(n+1)$-st point from the left. Prove that the distribution $2(\xi^*_{n+1} - \tfrac{1}{2})\sqrt{2n}$ converges to $N(0, 1)$ as $n \to \infty$.
Hint: Use the fact that the density of ξ^*_{n+1} has the form

$$C^n_{2n+1}(n+1)x^n(1-x)^n.$$

308. Suppose the random variable ξ_n has a β-distribution with parameters $np > 0$ and $nq > 0$. This means that its density is

$$\beta(x; np; nq) = \begin{cases} \dfrac{\Gamma(np + nq)}{\Gamma(np)(nq)}\, x^{np-1}(1-x)^{nq-1}, & \text{for} \quad x \in [0, 1], \\ 0 & \text{for} \quad x \notin [0, 1]. \end{cases}$$

Prove that

$$\sqrt{\frac{n}{pq}}\,(p+q)^{3/2}\left(\xi_n - \frac{p}{p+q}\right)$$

converges in distribution to $N(0, 1)$ as $n \to \infty$.

309. We say that the random variable ξ has a Γ-distribution with parameters $\alpha > 0$ and $\lambda > 0$ if its density is

$$f(x; \alpha, \lambda) = \begin{cases} \dfrac{\alpha^\nu}{\Gamma(\lambda)} x^{\lambda-1} e^{-\alpha x} & \text{for} \quad x > 0, \\ 0 & \text{for} \quad x \leqslant 0. \end{cases}$$

Let ξ_n be a random variable having a Γ-distribution, and let

$$M[\xi_n] = \frac{n}{\alpha}, \quad D[\xi_n] = \frac{n^2}{\alpha^2}.$$

Prove that the distribution $\sqrt{n}\,(\alpha\xi n/n - 1)$ converges for fixed α as $n \to \infty$ to $N(0, 1)$.

5.4 Properties of c.f.'s and g.f.'s

310. Prove that if $f(t)$ is a characteristic function and if $f(t)\,g(ht)$ is a characteristic function for each of an infinitely increasing sequence of values h, then $g(t)$ is a characteristic function.

311. Prove that an arbitrary characteristic function $f(t)$ is positive definite, i.e.,

$$\sum_{k, m=1}^{n} f(t_k - t_m)\, z_k \bar{z}_m \geqslant 0,$$

for arbitrary complex numbers z_k, real numbers t_k and natural number n. (This property is not only necessary but for $f(0)=1$ also a sufficient condition for the continuous function $f(t)$ to be a c.f.)

312. Prove that the function $f(t)=1-|t|/a$ for $|t|<a$, having period $2a$, is a c.f. Using the converse of the theorem proved in the preceding problem (see the remark in parentheses), one can prove a general theorem. If $f(t)$ is a c.f., equal to zero for $|t|>a$ and if $g(t)=f(t)$ for $|t|<a$ and $g(t+2a)=g(t)$, then $g(t)$ is also a c.f.

313. Prove that the following functions cannot be c.f.'s:

a) $e^{-|t|\,i}$; $\quad \dfrac{1}{1-|t|\,i}$;

b) real function not having the property of evenness;

c) $f(t) = \begin{cases} 1 - t^2 & \text{for } |t| < 1 \\ 0 & \text{for } |t| \geqslant 1; \end{cases}$

d) $f(t) = \cos(t^2)$

314. Prove that if $f(t)$ is a c.f. then the functions $g_1(t) = e^{f(t)-1}$ and $g_2(t) = (1/t)\int_0^t f(z)\,dz$ are also characteristic functions.

315. Prove that for a real c.f. the following inequalities are valid:

a) $1 - f(nt) \leqslant n^2 (1 - f(t))$, $n = 0, 1, 2, 3, \ldots$

b) $1 + f(2t) \geqslant 2\{f(t)\}^2$.

316. Prove the following properties of a c.f.:

a) $|f(t+h) - f(t)| \leqslant \sqrt{2[1 - Rf(h)]}$;

b) $1 - Rf(2t) \leqslant 4(1 - Rf(t))$,

where $Rf(t)$ is the real part of the c.f.

317. Show that $\phi_{\xi,\eta}(z) = \phi_\xi(z)\,\phi_\eta(z)$ does not imply that ξ and η are independent.

318. Prove that one can find independent random variables ξ_1, ξ_2, ξ_3 such that the distribution functions of ξ_1, ξ_2, and ξ_3 are distinct, but the c.f. of the distributions $\xi_1 + \xi_2$ and $\xi_2 + \xi_3$ coincide (see Problem 312).

319. Prove that $f_{\xi+\eta}(t) = f_\xi(t)\,f_\eta(t)$ does not imply that ξ and η are independent.

Hint: Look at the probability density $p(x, y) = \frac{1}{4}(1 + xy(x^2 - y^2))$ for $|x| < 1$, $|y| < 1$ and $p(x, y) = 0$ in the remaining cases.

320.* Prove that the c.f. of the random variable ξ is differentiable at zero does not imply that $M[\xi]$ exists.

5.5 Solution of problems with the aid of c.f.'s and g.f.'s

321. Solve Problem 129 using multidimensional characteristic functions. Show that if in performing an experiment, any one of n pairwise incompatible results A_k $(k = 1, 2, \ldots, n)$ are possible, $(P\{A_k\} = p_k; \sum p_k = 1)$ and this experiment is repeated v times, where v does not depend on which of the A_t occurred, then in the case when v has a Poisson distribution, η_k is the number of realizations, A_k $(k=1, 2, ..., n)$ are independent.

322. A point M moves along an integral line, passing in one step from the point n $(A<n<B)$ to the point $n+1$ with probability p and to the point $n-1$ with probability $1-p$. The motion begins from zero. Denote by τ the first moment the point is at A or B. Find the distributions of τ and $M[\tau]$. Also find the probability that the point first is at A.[1]

323. In an urn there were M red balls and $N-M$ white balls. The balls were taken out of the urn one after the other. Suppose the first red ball taken out appeared at the k_1-th removal, the second at the k_2-th, ..., M-th red ball at the k_M-th removal. Set $\xi = \sum k_i$. In order to find $P\{\xi=n\}$, we use the classical definition of probability. The number of all possible ways of taking balls out of the urn is C_N^M. Denote by $A_n(M, N)$ the number of them favorable with respect to the event $\xi=n$. Find $\sum_{k=0}^{N} \sum_n A_n(k, N) x^n y^k$. For $M=3$ and $N=3$, find $P\{\xi=n\}$. This problem is closely related to the nonparametric criterion of Wilcoxson.[2]

324. Use the c.f., show that, for $p<\frac{1}{2}$,

$$\max_{l} \left| \sum_{k=0}^{l} \left[C_n^k p^k (1-p)^{n-k} - \frac{(np)^k}{k!} e^{-np} \right] \right| \leqslant C_p,$$

where C is a constant not depending on n. (Another estimate of the closeness of the Poisson distribution to the binomial distribution can be found in Problem 246.)

325.* $2n$ points at equal distances are marked off on a circle. These points are randomly grouped in n pairs and the points of each of the pairs are connected by a chord. What is the probability that the n chords constructed do not intersect?

Hint: We use the fact that the number of "favorable" outcomes M_n satisfies the relation

$$M_n = \sum_{r=0}^{n-1} M_r \cdot M_{n-r-1}.$$

---

[1] This scheme was used by A. Wald to estimate the effectiveness of sequential analysis to differentiate two simple hypotheses.
[2] See the footnote to Problem 32.

326. Show that if $1-F(x)=o(e^{-cx})$ as $x\to\infty$ and $F(x)=o(e^{-c|x|})$ as $x\to-\infty$ $(c>0)$, then the distribution is uniquely defined by its moments.

327. A series of independent experiments is performed, for each of which the probability of a favorable outcome equals $p=1-q$, until v successive favorable outcomes are obtained where $v>0$ is given. Let p_{nv} denote the probability that to achieve this goal exactly n experiments are necessary. Prove that the generating function is

$$\phi(z)=\sum_{n=1}^{\infty}p_{nv}z^n=\frac{p^vz^v(1-pz)}{1-z+p^vqz^{v+1}}$$

and show that $M(n)=\phi'(1)=1-p^v/p^vq^v$.

328. n Bernoulli experiments are performed and it is noted that μ is the largest number of successive favorable outcomes occurring for these n experiments. Denoting by $P_{n,v}=P\{\mu\leqslant v\}$, show that

$$P_{n,v}=1-p_{1,v}-\cdots-p_{n,v},$$

where the $p_{n,v}$ are defined in the preceding problem, and, consequently, that

$$\psi(z)=\sum_{n=1}^{\infty}P_{n,v}z^n=\frac{1-\psi(z)}{1-z}=\frac{1-p^vz^v}{1-z+p^vqz^{v+1}}.$$

Prove that

$$M[\mu]=\frac{\log n}{\log\dfrac{1}{p}}+O(1),\qquad D[\mu]=O(1).$$

See page 137 for the answers on problems 281–327.

6 *Application of measure theory*

Let us recall the definitions and theorems which will be needed to solve problems of this chapter.

Measurability. If the family $\mathfrak{M}$ of subsets of the set $\Omega = \{\omega\}$ satisfies the following conditions:

1. If $A \in \mathfrak{M}$, then $\bar{A} \in \mathfrak{M}$.
2. If $A_i \in \mathfrak{M} (i = 1, 2, \ldots)$, then $\bigcup_{i=1}^{\infty} A_i \in \mathfrak{M}$ and $\bigcap_{i=1}^{\infty} A_i \in \mathfrak{M}$,

then we say that $\mathfrak{M}$ is a *σ-algebra* in the space Ω.

The function $f(\omega)$ is said to be *measurable with respect to the σ-algebra* $\mathfrak{M}$, or simply $\mathfrak{M}$-*measurable*, if for an arbitrary C,

$$\{\omega : f(\omega) < C\} \in \mathfrak{M}.$$

Various notions of convergence. A sequence of random variables $\xi_n (n = 1, 2, \ldots)$ *converges in probability* to the random variable ξ if, for an arbitrary $\varepsilon > 0$,

$$P\{|\xi_n - \xi| > \varepsilon\} \to 0 \quad (n \to \infty).$$

As was already pointed out in Chapter 4, convergence in probability is denoted by $\xi_n \overset{p}{\to} \xi$ or by $p \lim \xi_n = \xi$. If

$$P\{\omega : \xi_n(\omega) \to \xi (n \to \infty)\} = 1,$$

then we talk of *convergence with probability* 1; such convergence occurs if, and only if, for an arbitrary $\varepsilon > 0$,

$$\lim_{n \to \infty} P\{\sup_{m \geqslant n} |\xi_m(\omega) - \xi(\omega)| > \varepsilon\} = 0.$$

The sequence $\xi_n (n = 1, 2, \ldots)$ is said to converge to ξ *in the mean* if, for all n, $M[|\xi_n|^2] < \infty$, $M[|\xi|^2] < \infty$ and $M[|\xi_n - \xi|^2] \to 0 (n \to \infty)$. This convergence is denoted by the symbol l.i.m. $\xi_n = \xi$. Let $F_n(x)$ be a sequence of distribution functions. If for all x which are points of continuity of the distribution function $F(x)$, $F_n(x) \to F(x)$, then we say that *the sequence F_n converges weakly to $F(x)$*. We denote weak convergence by $F_n \overset{w}{\to} F$. F_n converges to F uniformly if $\sup_x |F_n(x) - F(x)| \to 0$ $(n \to \infty)$. In the case when $\int_{-\infty}^{\infty} |d(F_n - F)| \to 0 (n \to \infty)$, *convergence in variation* holds. If the terms of the sequence $\{F_n\}$ are distribution functions of random variables ξ_n and if this sequence converges weakly to $F(x)$, the distribution function of the random variable ξ, then we say that the sequence ξ_n converges to ξ in distribution. The interrelationships among the various notions of convergence are clarified in Problems 334–343.

In order to solve a number of problems the *Borel-Cantelli lemmas* are needed. Let $A_1, ..., A_n, ...$ be an infinite sequence of events and let $P\{A_k\} = p_k$.

LEMMA 1. *If $\sum p_k < \infty$, then only a finite number of events A_k occur with probability* 1.

LEMMA 2. *If the events A_k are mutually independent and the series $\sum p_k$ diverges, then an infinite number of events A_k occur with probability* 1.

As a rule, the solution of problems on the investigation of the convergence of series of random variables is based on the following *three series theorem*, due to A. N. Kolmogorov. *Let ξ_n be a sequence of independent random variables, C a positive constant, and $A_n = \{\omega : |\xi_n(\omega)| \leqslant C\}$. Then: a necessary and sufficient condition for the series $\sum_{n=1}^{\infty} \xi_n(\omega)$ to converge almost everywhere is that the following three series converge:*

1. $\sum_{n=1}^{\infty} P\{\bar{A}_n\}$;

2. $\sum_{n=1}^{\infty} M[\xi \chi_{A_n}(\omega)]$;

3. $\sum_{n=1}^{\infty} D[\xi \chi_{A_n}(\omega)]$, *where $\chi_{A_n}(\omega)$ is the characteristic function (indicator) of the set A_n.*

STRONG LAW OF LARGE NUMBERS (SLLN) (see §34 of thetextbook by B. V. Gnedenko). Let $\{\xi_n\} (n = 1, 2, ...)$ be a sequence of random variables. We say that it is subject to SLLN if, with probability 1,

$$\frac{1}{n} \sum_{k=1}^{n} \xi_k - \frac{1}{n} \sum_{k=1}^{n} M[\xi_k] \to 0.$$

THEOREM OF A. N. KOLMOGOROV. *To apply SLLN to the sequence of*

Application of measure theory

mutually independent variables $\{\xi_n\}$, it is sufficient that

$$\sum_{n=1}^{\infty} \frac{D[\xi_n]}{n^2} < \infty.$$

The proof of this theorem is based on a remarkable generalization of Chebyshev's inequality – the *Kolmogorov inequality*: for the independent $\xi_1, \xi_2, ..., \xi_n$, having finite dispersion, the probability of the simultaneous occurrence of the inequalities

$$\left| \sum_{i=1}^{k} (\xi_i - M[\xi_i]) \right| < \varepsilon \quad (k = 1, 2, ..., n)$$

is not less than

$$1 - \frac{1}{\varepsilon^2} \sum_{k=1}^{n} D[\xi_k].$$

Conditional probabilities and mathematical expectations. Let $(\Omega, \mathfrak{M}, P)$ be a given probability space, where $\Omega = \{\omega\}$ is a set, $\mathfrak{M}$ is a σ-algebra of subsets of Ω, and P is the probability measure on $\mathfrak{M}$. Let $\mathfrak{N}$ be a σ-algebra in Ω, where $\mathfrak{N} \leqslant \mathfrak{M}$, and let $\xi(\omega)$ be a summable function on Ω. Any $\mathfrak{N}$-measurable function satisfying for an arbitrary $A \in \mathfrak{N}$ the relation

$$\int_A \xi(\omega) P(d\omega) = \int_A M(\xi \mid \mathfrak{N}) P(d\omega)$$

is called the *conditional mathematical expectation $M(\xi \mid \mathfrak{N})$ of ξ with respect to $\mathfrak{N}$.* With the aid of the Radon-Nikodým theorems one can show that such a function always exists. It is easy to see that the function $M[\xi \mid \mathfrak{N}]$ is defined only to within an arbitrary set of the σ-algebra $\mathfrak{N}$ having P-measure zero. The function $M(\chi_B \mid \mathfrak{N})$ is called the conditional probability of the event B with respect to the σ-algebra $\mathfrak{N}$. It can also be defined as an $\mathfrak{N}$-measurable function which satisfies, for an arbitrary $A \in \mathfrak{N}$, the relation

$$P(A \cap B) = \int_A P(B \mid \mathfrak{N}) P(d\omega).$$

The fundamental properties of a conditional mathematical expectation are clarified in Problems 364–367.

For additional reading see [9], [10], [12] and [13].

6.1 Measurability

329. Prove that every finite σ-algebra $\mathfrak{M}$ in the space $\Omega = \{\omega\}$ is connected with some decomposition of this space into a finite number of disjoint sets $A_1, A_2, \ldots, A_n$ in the following way: $\mathfrak{M}$ consists of all possible finite unions of the sets A_i. Prove also that the function $f(\omega)$ is measurable with resect to the σ-algebra $\mathfrak{M}$ if, and only if, it takes on constant values on each of the A_i.

330. Consider the Hilbert space L^2 of measurable functions which are square integrable and let H be a subspace of the space L^2 consisting of the functions which are measurable with respect to a finite σ-algebra $\mathfrak{M}$.

Find the dimension of H. Prove that $f \perp H$ if, and only if, the integral of f over an arbitrary set $C \in \mathfrak{M}$ equals zero.

331. *Example of a non-measurable set.* Let us assume that in the unit square there is located a point and let $0 \leqslant \xi, \eta \leqslant 1$ be its coordinates. We introduce $\Omega = \{\omega\}$, the space of elementary events. For an arbitrary subset A of the interval $[0, 1]$ we set $\hat{A} = \{\omega : \xi(\omega) \in A\}$. Let $\mathfrak{M}$ be a σ-algebra of the sets $\hat{A}$, corresponding to the sets A which are measurable in the sense of Lebesgue. We define on Ω a probability measure by setting $\mu\{A\}$ equal to the Lebesgue measure of the set A. The set

$$M = \{\omega; \eta(\omega) \leqslant \tfrac{1}{2}\}$$

will not be measurable with respect to the σ-algebra $\mathfrak{M}$.

Show that

$$\mu^*\{M\} \equiv \inf\left\{ \sum_{n=1}^{\infty} \mu\{\hat{A}_n\}; \hat{A}_n \in \mathfrak{M}, \quad n = 1, 2, \ldots, M \subset \bigcup_n \hat{A}_n \right\} = 1,$$

and

$$\mu_*\{M\} \equiv \sup\{\mu\{\hat{A}\} : M \supset \hat{A} \in \mathfrak{M}\} = 0.$$

332. Let $\xi_k (k = 1, 2, 3, \ldots)$ be a sequence of mutually independent random variables with the same distributions. Assume that ξ_k does not have a finite mathematical expectation and let A be an arbitrary positive constant. Show that the event C, defined by the requirement that for infinitely many n the events $|\xi_n| > A_n$ are realized, is measurable. Find $P\{C\}$.

333. Let $\xi_1, \xi_2, \ldots$ be a sequence of independent random variables, each of which takes on only two values: 0 and 1. Moreover, $P\{\xi_n = 1\} = p_n$ and

$$\sum_{n=1}^{\infty} p_n < \infty.$$

Set $\zeta_n = \sum_{i=1}^{n} \xi_i$. Prove that for an arbitrary $k \geq 0$ the sets of elementary events $A_k = \{\omega : \zeta_n(\omega) \to k (n \to \infty)\}$ are measurable. Prove that $\sum P\{A_k\} = 1$. i.e., that the sequence ζ_n is bounded with probability 1.

6.2 Various concepts of convergence

334*. Let $\xi_i (i = 1, 2, \ldots, n)$ be independent, identically distributed random variables, where $M[\xi] = a$ and $D[\xi] = \sigma^2 > 0$. Find the limit distribution as $n \to \infty$ for

$$\zeta_n = \frac{1}{\sigma \sqrt{n}} \left(\sum_{i=1}^{n} \xi_i - na \right)$$

and prove that there does not exist a random variable such that

$$\lim_{n \to \infty} P\{|\zeta_n - \zeta| < \varepsilon\} = 1.$$

Remark. It follows from this problem, in particular, that the convergence of the distributions does not imply the convergence in probability of the corresponding random variables.

335. Prove that convergence with probability 1 implies convergence in probability.

336. Prove that $\xi_n \xrightarrow{P} \xi$ if, and only if, every subsequence $\{\xi_{n_i}\}$ of the sequence ξ_n contains another subsequence which converges to ξ with probability 1.

337. Prove that convergence in mean implies convergence in probability.

338. Construct an example of a sequence which converges in the mean, but does not converge with probability 1.

339. Construct an example of a sequence which converges with probability 1, but does not converge in the mean.

340. Prove that if $F(x)$ is continuous, then the weak convergence of $F_n(x)$ to F implies that F_n converges uniformly to F.

341. Show that the weak convergence of distribution functions does not imply convergence in variation.

342. Show that the convergence in variation of distribution functions implies uniform convergence, and that uniform convergence, in turn, implies the weak convergence of distribution functions.

343. Prove that if F_n are distribution functions of integer-valued random variables, then the weak convergence of F_n to F implies convergence in variation.

6.3   Series of independent random variables

344. Let ξ_n be a sequence of random variables. Prove that the event C, which is defined by the convergence of the series

$$\sum_{i=1}^{\infty} \xi_i,$$

is measurable.

345. Suppose given a sequence $\xi_1, \xi_2, \ldots$ of independent random variables with arbitrary distribution functions (for simplicity we can assume that these functions are the same for all ξ_i). It is known that $\sigma_n^2 = M[\xi_n^2] < \sigma^2$ and $M[\xi_n] = 0$. Prove that the series $\sum_{n=1}^{\infty} \xi_n/2^n$ converges with probability 1.

346. A point ξ is located at random on the segment $[0, 1]$. Define a function $\phi_n(\xi)$ by setting $\phi_n(\xi) = +1$ or -1 depending on whether the positive integer i for which $(i-1)/2^n \leqslant \xi < i/2^n$ is odd or even. Prove that the series

$$\sum_{k=1}^{\infty} C_k \phi_k(\xi)$$

converges with probability 1 if, and only if, the series $\sum_{k=1}^{\infty} C_k^2$ converges.

347. Show that, independently of the choice of the C_k in the preceding problem, the event C that the series

$$\sum_{k=1}^{\infty} C_k \phi_k(\xi)$$

converges, is, for an arbitrary n, measurable with respect to $\phi_n(\xi)$, $\phi_{n+1}(\xi), \ldots$. Prove that for arbitrary $\varepsilon > 0$, and n, there exist a number $N > n$ and an event A, measurable with respect to $\phi_n(\xi)$, $\phi_{n+1}(\xi), \ldots$, $\phi_N(\xi)$, such that $P\{C \triangle A\} < \varepsilon$.

348. Prove that if the random variables $\xi_1, \xi_2, \ldots$ are independent and the event C is, for arbitrary n, measurable with respect to $\xi_n, \xi_{n+1}, \ldots$, then $P\{C\} = 0$ or 1 (*A. N. Kolmogorov's Zero-One Law*).
Hint. Use the fact that, for an arbitrary event A, depending only on a finite number of ξ_i's,

$$P\{C \cap A\} = P\{C\} P\{A\}.$$

349. Let $\{A_k\}$ $(k = 1, 2, \ldots)$ be a sequence of mutually independent events; then, with probability 1 there is realized a finite or infinite number of A_k depending on whether the series

$$\sum_{k=1}^{\infty} P\{A_k\}$$

converges or diverges (Borel-Cantelli lemma). Prove this result using the three series theorem.

350. Let $\{\xi_n\}$ $(n = 1, 2, \ldots)$ be a sequence of random variables, each having a mathematical expectation, and ξ is a random variable, having a dispersion, such that for an arbitrary positive integer n, the functions $\xi_1, \ldots, \xi_n$, $\xi - (\xi_1 + \cdots + \xi_n)$ are independent. Prove that in this case all the ξ_n have a dispersion, and the series

$$\sum_{k=1}^{\infty} (\xi_k - M[\xi_k])$$

converges almost everywhere.

351.* Let $\bar{\alpha} = [0, \alpha]$, $\bar{\beta} = [\alpha, \beta]$ and $\overline{\triangle} = [0, \triangle]$ be given segments, where $\triangle < -[\alpha, \beta]$. On one of the segments $\bar{\alpha}$ or $\bar{\beta}$ a point x_0 is taken which then moves according to the following law: if at the moment n the point is at the position x_n on the segment $\bar{\alpha}$ (respectively $\bar{\beta}$), then at the moment $n+1$ it goes to the position $x_{n+1} = x_n + \triangle$ on the same segment in the case when $\alpha - x_n \geqslant \triangle$ (respectively $\beta - x_n \geqslant \triangle$). But if the opposite inequalities are satisfied, then the point x_{n+1} turns out to be with probability $\frac{1}{2}$ at a distance $\triangle' = \triangle - (\alpha - x_n)$ (respectively $\triangle'' = \triangle - (\beta - x_n)$) right from the left end of the segment on which it was earlier, and with probability $\frac{1}{2}$ it will be at the same distance right from the left end of the other segment. Prove that if α, $\alpha - \beta$ and $\triangle$ are incommensurable, then for an arbitrary initial position of the point and for an arbitrary interval of length γ, lying on one of the segments $\bar{\alpha}$ or $\bar{\beta}$, with probability 1 a number n can be found such that at the n-th step the point falls in the interior of the interval γ.

352. In a sequence of Bernoulli trials, let A_n be the event that a series of n successive successes occurs between the 2^n-th and the 2^{n+1}-st trials. Prove that if $p \geqslant \frac{1}{2}$, then with probability 1 infinitely many events A_n are realized; if $p < \frac{1}{2}$, then with probability 1 only a finite number of events A_n are realized.

6.4 Strong law of large numbers and the iterated logarithm law

353. Show that whenever the sequence of nonnegative numbers $\{\sigma_n^2\}$ is such that

$$\sum_{n=1}^{\infty} \frac{\sigma_n^2}{n^2} = \infty,$$

there exists a sequence of independent random variables $\{\xi_n\}$ such that $M[\xi_n] = 0$, $D[\xi_n] = \sigma_n^2 \ (n = 1, 2, \ldots)$ and

$$\left\{ \frac{1}{n} \sum_{k=1}^{n} \xi_k \right\}$$

does not converge to zero almost everywhere.

354. Two sequences of random variables $\{\xi_n\}$ and $\{\eta_n\}$ are called *equivalent* if

$$\sum_{n=1}^{\infty} P\{\xi_n \neq \eta_n\} < \infty .$$

Prove that if the sequence of independent random variables $\{\xi_n\}$ is subject to the SLLN, then there exists a sequence $\{\eta_n\}$ equivalent to it such that

$$\sum_{n=1}^{\infty} \frac{D[\eta_n]}{n^2} = \infty .$$

In other words, prove that the proposition converse to Kolmogorov's theorem on the SLLN does not hold.

355. Show that there exists the following somewhat weakened converse of the theorem on the SLLN. If $\{\xi_n\}$ is a sequence of independent random variables such that $M[\xi_n] = 0$, $|(1/n)\xi_n| \leqslant C$, $n = 1, 2, \ldots$, where C is some constant, and with probability 1 the sequence

$$\left\{ \frac{1}{n} \sum_{i=1}^{n} \xi_i \right\}$$

converges to zero, then $\sum D[\xi_n]/n^{1+\varepsilon} < \infty$ for arbitrary $\varepsilon > 0$.
Hint. Use the fact that if the sequence of real numbers $\{y_n\}$ is such that the sequence

$$\left\{ \frac{1}{n} \sum_{i=1}^{n} y_i \right\}$$

converges to zero or is at least bounded, then the series

$$\sum_{n=1}^{\infty} \frac{y_n}{n^{1+\varepsilon}}$$

converges for arbitrary positive ε.

356. Example of an unfavorable fair game. Suppose the possible values of winning at each trial will be $0, 2, 2^2, 2^3, \ldots$; the probability that the win equals 2^k is

$$p_k = \frac{1}{2^k k(k+1)},$$

and that the probability of the zero winning equals $p_0 = 1 - (p_1 + p_2 + \cdots)$. Then the expected amount won per trial is

$$\mu = \sum 2^k p_k = (1 - \tfrac{1}{2}) + (\tfrac{1}{2} - \tfrac{1}{3}) + (\tfrac{1}{3} - \tfrac{1}{4}) + \cdots = 1.$$

Assume that the player at every trial, for his right of participation in the game, pays a ruble so that after n trials his pure gain (or loss) equals $S_n - n$, where S_n is the sum of n independent random variables with the distribution given above. Show that for every $\varepsilon > 0$ the probability that in n trials the player loses more than $(1-\varepsilon)n/\log_2 n$ rubles, tends to 1, where $\log_2 n$ denotes the logarithm to the base 2. In other words, it must be proved that

$$P\left\{S_n - n < -\frac{(1-\varepsilon)n}{\log_2 n}\right\} \to 1.$$

Hint. Use the "method of truncation" with limit $n/\log_2 n$. Show that for the "truncated" quantities $\bar{\xi}_k$,

$$P\left\{|\bar{\xi}_1 + \cdots + \bar{\xi}_n - nM[\bar{\xi}_1]| < \frac{\varepsilon n}{\log_2 n}\right\} \to 1,$$

$$1 - \frac{1}{\log_2 n} \geq M[\bar{\xi}_1] \geq 1 - \frac{1+\varepsilon}{\log_2 n}.$$

357. Let $\{\xi_n\}$ be a sequence of mutually independent random variables such that $\xi_n = \pm 1$ with probability $(1 - 2^{-n})/2$ and $\xi_n = \pm 2^n$ with proba-

91

bility 2^{-n-1}. Prove that the law of large numbers is applicable to $\{\xi_n\}$. *Remark.* This means that the condition

$$\frac{1}{n} D[\xi_1 + \cdots + \xi_n] \to 0$$

is not necessary.

358. Prove that if $\xi_1, \xi_2, \ldots$ are pairwise independent, identically distributed and have finite dispersion, then the SLLN is applicable.

359. Let $\xi_k (k=1, 2, \ldots)$ be independent,

$$M[\xi_k] = m_k > 0 ; \qquad \sum_{k=1}^{\infty} m_k = \infty \quad \text{and} \quad \sum_k \frac{D[\xi_k]}{\left(\sum\limits_{i=1}^{k} m_i\right)^2} < \infty .$$

Prove that in this case,

$$\lim_{n \to \infty} \frac{\sum\limits_{k=1}^{n} \xi_k}{\sum\limits_{k=1}^{n} m_k} = 1$$

with probability 1.

360. Suppose the random variables $\xi_1, \xi_2, \ldots$ are independent and have a normal distribution with mathematical expectation 0 and dispersion 1, and suppose the event A_n represents $S_n = \xi_1 + \xi_2 + \cdots + \xi_n > \lambda(2n \log \log n)^{1/2}$, where λ is a fixed constant. Prove[1]) that, for $\lambda > 1$, only a finite number of A_n's are realized with probability 1.
Hint. Suppose the event B_r consists of the following: that in the interval of time from γ^r to γ^{r+1}, where $1 < \gamma < \lambda$, A_n is realized at least once. Use the fact that if infinitely many events A_n are realized, then also infinitely many events B_r are realized.

361. Suppose the event A_n is defined as in the preceding problem. Prove

---

[1]) The propositions proved in Problems 360–361 form a particular case of the so-called *iterated logarithm law*. For the Bernoulli scheme this law was first obtained by A. Khinchin (see Fundamenta Mathematicae 6, 9–20, 1924). Later, a more general theorem was proved by A. N. Kolmogorov (see "Mathematische Annalen", 101, 126–135, 1929).

that, for $\lambda < 1$, infinitely many events A_n are realized with probability 1. *Hint.* Let γ be so large that $1 - \gamma^{-1} > \lambda$, and let n_r be the nearest integer in γ^r. Set $D_r = S_{n_r} - S_{n_{r-1}}$. We shall first prove that for infinitely many r,

$$D_r > \lambda (2n_r \log \log n_r)^{1/2} .$$

362. Let S_n be the number of successes with n Bernoulli trials with probability of success p. Prove that

$$\limsup_{n \to \infty} \frac{S_n}{(2np(1-p) \log \log n)^{1/2}} = 1$$

(see Problems 360–361).

6.5 Conditional probabilities and conditional mathematical expectations

363. Let $A_1, A_2, \ldots, A_n$ be a complete system of events, i.e., $A_i \cap A_j = \phi$ $(i \neq j)$ and

$$\bigcup_{i=1}^{n} A_i = \Omega ,$$

and $B_1, B_2, \ldots, B_m$ another complete system. Prove the following generalized total probability formula for an arbitrary event C,

$$P\{C\} = \sum_{i=1}^{n} \left[\sum_{j=1}^{m} P\{C \mid A_i \cap B_j\} P\{B_j \mid A_i\} \right] P\{A_i\} .$$

364. Show that if $\mathfrak{A}$ is an algebra consisting of exactly two elements: ϕ and the entire space Ω, then for an arbitrary random variable ξ, having expectation

$$M[\xi \mid \mathfrak{A}] = M[\xi] .$$

365. Find $M[\xi \mid \mathfrak{A}]$ as a function of ω in the case when the σ-algebra consists of exactly 4 elements $\{\phi, A, \bar{A}, \Omega\}$, where $0 < P\{A\} < 1$. Into what do the formulas obtained go when $\xi(\omega) = \chi_B(\omega)$, where χ_B is the c.f. of the set B, introduced in Problem 10?

366. Prove that if the random variable ξ is measurable relative to $\mathfrak{A}$ and bounded, then

$$M[\xi \mid \mathfrak{A}] = \xi$$

almost everywhere on Ω.

367. Let $(\Omega, \mathfrak{M}, P)$ be a given probability space and let $\mathfrak{A} \subseteq \mathfrak{M}$ be a σ-algebra. Prove:

a) if $\xi \geqslant \eta$ and $M[\xi]$ and $M[\eta]$ exist, then almost everywhere on Ω,

$$M[\xi \mid \mathfrak{A}] \geqslant M[\eta \mid \mathfrak{A}];$$

b) if $M[\xi]$ and $M[\eta]$ exist, then

$$M[a\xi + b\eta \mid \mathfrak{A}] = aM[\xi \mid \mathfrak{A}] + bM[\eta \mid \mathfrak{A}];$$

c) if $M[\xi]$ exist and $0 \leqslant \xi_n \uparrow \xi$, then

$$M[\xi_n \mid \mathfrak{A}] \uparrow M[\xi \mid \mathfrak{A}];$$

d) if $M[\xi]$ and $M[\eta]$ exist, and ξ is measurable relative to $\mathfrak{A}$, then

$$M[\xi\eta \mid \mathfrak{A}] = \xi M[\eta \mid \mathfrak{A}].$$

368. A random point occurs with a uniform distribution on the segment $[-a, a]$. Two pieces of apparatus observe it. The first apparatus registers it at the point x with probability $p(x)$ and the second with the probability $q(x)$. Find the probability that the point, if observed by both pieces of apparatus, occurs to the left of x.

369. Two points are chosen at random in the interior of the unit circle independently one of the other. Considering all positions of the points equally possible, construct the function of the distribution of the distances between the points.

370. Let $\xi_i (i=1, 2, ..., n)$ be independent normally distributed random variables with parameters $(0, 1)$. Find the expection and density of the distribution of $\sum_{i+1}^{n} \xi_i$ under the condition that $\sum_{i+1}^{n} \xi_i = a$.

371. Let $\xi_i (i=1, 2, ..., n)$ be random variables having joint distribution density $h(x_1, ..., x_n)$ and let $0 \leqslant \eta = f(\xi_1, \xi_2, ..., \xi_n) \leqslant C < \infty$. Find:

a) the expectation of the random variable η under the condition that

$$\sum \xi_i^2 = p^2 \, ;$$

b) the density of the distribution of η under the same condition.

372. Let us assume that the random variable ξ has density of distribution $f(x, a)$, depending on the unknown parameter a. There were obtained n independent realizations of $\xi : x_1, ..., x_n$. Find the *a posteriori* density of the distribution of the parameter a, $\phi(a \mid x_1, ..., x_n)$, if the *a priori* density of the distribution a, $\phi_1(a)$, is known.

See page 139 for the answers on problems 329–371.

7

Infinitely divisible distributions. Normal law. Multidimensional distributions

The only new concept in this chapter is that of the infinitely divisible (i.d.) distribution law; in this connection, see Chapter 9 of the textbook by B. V. GNEDENKO. The distribution law $F(x)$ is called i.d. if its characteristic function, for an arbitrary integer $n \geqslant 1$, can be written in the form

$$f(t) = [f_n(t)]^n,$$

where $f_n(t)$ is also a characteristic function. In Problems 375, 381–387, it is assumed that the general form of the logarithm of the characteristic function of the i.d. law

$$\log f(t) = i\gamma t + \int\limits_{-\infty}^{\infty} \left(e^{itu} - 1 - \frac{itu}{1 + u^2} \right) \frac{1 + u^2}{u^2} \, dG(u), \tag{1}$$

is known, where $G(u)$ is a nondecreasing function of bounded variation, and the function under the integral sign is defined by the equality

$$\left[\left\{ e^{itu} - 1 - \frac{itu}{1 + u^2} \right\} \frac{1 + u^2}{u^2} \right]_{u=0} = -\frac{t^2}{2}$$

for $u=0$. It is also assumed known that the representation of $\log f(t)$ by formula (1) is unique.

We note that the material of the majority of the problems on the i.d. law is taken from the 2-nd and 3-rd chapters of the book by B. V. GNEDENKO and A. N. KOLMOGOROV. *Limiting distributions for sums of independent random variables.* GITTL, Moscow–Leningrad (1949). Also see [9] and [12].

7.1   Infinitely divisible distributions

373. Prove that if a characteristic function $f(t)$ is such that for two incommensurable values of the argument t_0 and t_1 the equalities $|f(t_0)| = 1$ and $|f(t_1)| = 1$ hold, then $|f(t)| \equiv 1$. What can be said about the corresponding distribution functions? Will they be infinitely divisible?

374. Prove that a random variable, distributed according to the Cauchy law

$$F(x) = \frac{1}{\pi}\left(\frac{\pi}{2} + \arctan\frac{x-b}{a}\right),$$

is infinitely divisible.

375. Prove that if the sum of two independent infinitely divisible random variables is distributed: a) according to Poisson's law; b) according to the normal law, then each term is distributed in the case a) according to the Poisson law; in case b) according to the normal law (also see Problem 396).

376. Prove that a random variable with density distribution

$$p(x) = \begin{cases} 0 & \text{for } x \leqslant 0; \\ \dfrac{\beta^\alpha}{\Gamma(\alpha)} x^{\alpha-1} e^{-\beta x} & \text{for } x > 0, \end{cases}$$

where $\alpha > 0$, $\beta > 0$ are constants, is infinitely divisible.
Remark. From this it follows that Maxwell's distribution and the χ^2 distributions are infinitely divisible.

377.* Prove that the characteristic functions of infinitely divisible laws do not vanish for $|t| < \infty$.
Hint. Use the inequality of Problem 316 b).

378. Prove that a distribution function which is a limiting function in the sense of weak convergence for an infinitely divisible law is infinitely divisible.

379. Leaning on the statement of the preceding problem, prove that if $f(t)$ is the characteristic function of an infinitely divisible law of distribution, then, for an arbitrary $c > 0$, the function $[f(t)]^c$ also is the characteristic function of an infinitely divisible law.

380. Prove that the collection of infinitely divisible laws coincides with the collection of laws which are the compositions of a finite number of

Poisson laws and limiting laws for them in the sense of weak convergence.

Hint. Use the following relation for the characteristic function: for an arbitrary $a \neq 0$, as $n \to \infty$,

$$n\left(\sqrt[n]{a} - 1\right) \to \log a.$$

381. Using formula (1) (see page 96), show that the characteristic function

$$f(t) = \frac{1-\beta}{1+\alpha} \frac{1+\alpha e^{-it}}{1-\beta e^{it}} \quad (0 < \alpha \leqslant \beta < 1)$$

is not infinitely divisible.

382. Prove that $|f(t)|$, where $f(t)$ is the characteristic function defined in the preceding problem, is infinitely divisible.[1]

383. Prove that: a necessary and sufficient condition for $F(x)$ in formula (1) (see page 96) to be integral is that the function $G(u)$ grows only at integral points $\neq 0$.

The assertion of this problem can be significantly strengthened, namely, see the following problem.

384.* Prove that there exists an absolute constant C such that for an arbitrary infinitely divisible law F, for which

$$\sum_{k=-\infty}^{\infty} \{F(k+1) - F(k+0)\} \leqslant \varepsilon,$$

the inequality

$$\int \left(\frac{1+u^2}{u^2}\right) dG(u) < C_\varepsilon$$

is valid, where the integral is taken over all integer points of the infinite line (Yu. V. Prokhorov).

385.* Prove that a non-degenerate infinitely divisible distribution cannot be concentrated on a finite interval.

---

[1] It follows from Problems 381–382 that an infinitely divisible characteristic function $[f(t)]^2$ decomposes into two non-infinitely divisible characteristic functions $f(t)$ and $\overline{f(t)}$.

386. Prove that if there exist $k>2$ first semi-invariants of an infinitely divisible distribution $K_1, K_2, \ldots, K_k$, then the sequence $\alpha_0 = K_2, \alpha_1 = K_3, \ldots,$ $\alpha_{k-2} = K_k$ is non-negative definite, i.e., for an arbitrary polynomial of degree not greater than $k-2$,

$$\mathscr{P}(x) = \sum_{u=0}^{k-2} A_u x^u \not\equiv 0, \, \mathscr{P}(x) \geqslant 0,$$

the functional

$$Q(\mathscr{P}) = \sum_{u=0}^{k-2} A_u \alpha_u \geqslant 0.$$

387.** We say that the infinitely divisible law has a bounded spectrum M if $G(-\infty) = G(-M)$ and $G(M) = G(\infty)$.

Let $\mathfrak{M}(M)$ be the collection of infinitely divisible distribution laws having a bounded spectrum M and dispersion equal to 1, let $\mathfrak{A}(l)$ be the collection of distribution functions of the random variables ζ such that $|\zeta| \leqslant l$ and $D[\zeta] = 1$, and let

$$\psi(n, l) = \inf_{F \in \mathfrak{A}(l)} \, \inf_{G \in \mathfrak{M}(M)} \, \sup_x |F^n(x) - G(x)|,$$

where

$$F^n(x) = \underbrace{F * \cdots * F(x)}_{n \text{ times}}.$$

Prove that for arbitrary $M, l < \infty$, there exists a $k < \infty$ such that for all sufficiently large n,

$$\psi(n, l) > n^{-k}.$$

388.* Domain of attraction; application of the Poisson law.
Let

$$S_n = \sum_{k=1}^{n} \xi_{k,n},$$

where $\xi_{k,n} \, (k = 1, 2, \ldots, n)$ are mutually independent random variables which take the value 1 with probability $p_{k,n}$ and 0 with probability $1 - p_{k,n}, \, 0 < p_{k,n} < 1$,

$$a_n = \sum_{k=1}^{n} p_{k,n}, \qquad b_n = \max_{1 \leqslant k \leqslant n} p_{k,n}.$$

Prove that a necessary and sufficient condition for the convergence of the distribution of the sums S_n to the Poisson law with parameter $0 < a < \infty$ is that the following conditions be satisfied:

$$\lim_{n \to \infty} b_n = 0 ; \qquad \lim_{n \to \infty} a_n = a .$$

7.2   The normal distribution

389.* A point $\boldsymbol{\xi} = (\xi_1, \ldots, \xi_n)$ is located at random on the n-dimensional sphere. Prove that for large values of n, ξ_1 will be distributed approximately normally.

390. Prove that for $x > 0$,

$$1 - \Phi(x) \approx \frac{1}{\sqrt{2\pi}} e^{-x^2/2} \times$$

$$\times \left\{ \frac{1}{x} - \frac{1}{x^3} + \frac{1 \cdot 3}{x^5} - \cdots + (-1)^k \frac{1 \cdot 3 \cdots \cdot (2k-1)}{x^{2k+1}} \right\},$$

where

$$\Phi(x) = \frac{1}{\sqrt{2\pi}} \cdot \int_{-\infty}^{x} e^{-u^2/2} \, du ;$$

moreover, for k even the right member exceeds $1 - \Phi(x)$, and for k odd it is less than $1 - \Phi(x)$.

391. Prove that for an arbitrary constant $a > 0$,

$$\left\{ 1 - \Phi\left(x + \frac{a}{x} \right) \right\} \Big/ \{1 - \Phi(x)\} \to e^{-a}$$

as $x \to \infty$.

392.* Construct an example showing from the conditions:
  a) ξ is normally distributed;
  b) η is normally distributed;
  c) $\operatorname{cov}(\xi, \eta) = 0$
it does not follow that ξ and η are independent.

393. Is a linear combination of normally distributed random variables always distributed normally?

394. Prove that if ξ and η are independent and normally distributed with parameters $a_1 = a_2 = 0$, $\sigma_1 = \sigma_2 = \sigma$, then the quantities $\rho^2 = \xi^2 + \eta^2$ and $\delta = \xi/\eta$ are also independent.

The following problem shows that the converse assertion is also valid.

395.* If ξ and η are independent and identically distributed, have density, and $\rho = \sqrt{\xi^2 + \eta^2}$ and $\delta = \xi/\eta$ are independent, then ξ and η are normally distributed.

396.** Let $\xi_1, \xi_2, \ldots, \xi_n$ be independent and their sum normally distributed. Prove that each of the ξ_i has a normal distribution.

397.* If $\xi_1, \xi_2, \ldots, \xi_n$ are independent, identically distributed and have dispersion, and if, moreover, an orthogonal matrix $C = \|c_{ij}\|$, different from the identity matrix, can be found such that the

$$\eta_j = \sum_{k=1}^{n} c_{jk}\xi_k$$

are independent, then the ξ_k are normally distributed. Consider the special case $n = 2$ and the matrix

$$C = \begin{pmatrix} \dfrac{1}{\sqrt{2}} & -\dfrac{1}{\sqrt{2}} \\ \dfrac{1}{\sqrt{2}} & \dfrac{1}{\sqrt{2}} \end{pmatrix}.$$

398. Let $p(x)$ be the density of a distribution. Then

$$H(p(x)) = - \int_{-\infty}^{\infty} p(x) \ln p(x)\, dx$$

is called the *entropy* of the continuous distribution. Prove that if

$\int_{-\infty}^{\infty} xp(x)dx=0$, $\int_{-\infty}^{\infty} x^2 p(x)\,dx=1$, then $H(p(x))\leqslant\ln\sqrt{2\pi e}$, where equality is attained only for a normal distribution.[1]

399. Let $p(x)=0$ for $x\leqslant 0$ and

$$\int_0^\infty xp(x)\,dx = a.$$

Prove that the maximum of the entropy holds when

$$p(x) = \frac{1}{a}\exp\left\{-\frac{x}{a}\right\},$$

and it equals $1n\ ea$.

400. The entropy of a multidimensional continuous distribution with density $P(x_1, x_2, ..., x_n)$ is defined by the formula

$$H = -\int ... \int p(x_1, ..., x_n)\ln p(x_1, ..., x_n)\,dx_1 ... dx_n.$$

Prove that for a multidimensional normal distribution with density

$$p(x_1, ..., x_n) = \frac{|a_{ij}|^{1/2}}{(2\pi)^{n/2}}\exp\left(-\tfrac{1}{2}\sum a_{ij}x_i x_j\right),$$
$$H = \ln(2\pi e)^{n/2}|a_{ij}|^{-1/2},$$

where $|a_{ij}|$ is the determinant whose elements are a_{ij}.

401. Let $p(x)$ be the density of the distribution ξ. Show that for an arbitrary n,

$$\prod_{k=1}^{n} p(x_k - x_0)$$

attains its maximum for

$$x_0 = \frac{1}{n}\sum_{k=1}^{n} x_k$$

---

[1] The entropy $H(p(x))$ was first introduced by C. Shannon. Yu. V. Linnik used it to prove the central limit theorem under the Lindeberg conditions (*Theory of probability and its applications*, IV, issue 3 (1959)).

if, and only if,

$$p(x) = \frac{1}{\sigma\sqrt{2\pi}} \exp\left\{-\frac{(x-a)^2}{2\sigma^2}\right\}.$$

402. Let $\mathfrak{A}$ be a family of distribution functions $F(x)$, defined by the following three conditions:

1) every distribution $F \in \mathfrak{A}$ is uniquely defined by its mathematical expectation μ and dispersion σ;

2) for arbitrary $a > 0$ and b, $F(x) \in \mathfrak{A}$ implies that $F(ax+b) \in \mathfrak{A}$;

3) $F \in \mathfrak{A}$ and $G \in \mathfrak{A}$ imply that $F * G \in \mathfrak{A}$. Prove that $\mathfrak{A}$ is the family of normal distributions.

7.3   Multidimensional distributions

403. Let ζ be a given random variable which takes on the complex values: $\zeta = \xi + i\eta$, where ξ, η are random variables which take on real values. Let $\zeta_t = \zeta e^{it}$ be a new random variable. What must the distribution ζ satisfy in order that all the variables ζ_t have the same distribution? One can assume that the random vector (ξ, η) has density.

404.* Let D be a simply connected region in the plane having a sufficiently smooth boundary and a density $p(x, y) > 0$ given in it. Prove that one can introduce new coordinates which will be independent.

405. Let ξ and η be two given random variables which take on the values:

$$\xi = a_1, a_2, \ldots, a_n;$$
$$\eta = b_1, b_2, \ldots, b_m.$$

Its joint distribution is given:

$$P\{\xi = a_i, \eta = b_j\} = p_{ij}.$$

A new random variable $\zeta = \phi(\xi, \eta)$ is defined where $\phi(x, y)$ is an arbitrary function of two variables on the set of pairs (a_i, b_j). Clearly, $M[\zeta^2] < \infty$, i.e.

$$\sum \phi^2(a_i b_j) p_{ij} < \infty.$$

If in the space of the functions ϕ, where ϕ is defined by the collection

103

of numbers:

$$\{c(i, j)\} \ (i = 1, ..., n; j = 1, ..., m),$$

one introduces a scalar product in the following way:

$$(\zeta_{\phi_1}\zeta_{\phi_2}) = M[\zeta_{\phi_1}\zeta_{\phi_2}] = \sum_{i, j} c_1(i, j) c_2(i, j) p_{ij};$$

then we obtain an Euclidean space $H_{\xi, \eta}$.

Consider the subspace H as a subspace of $H_{\xi, \eta}$, spanning the functions depending only on ξ,

$$H_{\xi} = \{\zeta_{\phi} : \phi(x, y) = \phi(x)\}.$$

Find the orthogonal projection η on the subspace H_{ξ}.

406. Let $\xi = \{\xi_1, ..., \xi_N\}$ be a random variable distributed in N-dimensional space, let $\mathbf{t} = \{t_1, ..., t_N\}$ be a vector of N-dimensional space, let

$$(\mathbf{t}, \xi) = \sum_{k=1}^{N} \xi_k t_k$$

be the scalar product. $\chi_{\xi}(\mathbf{t}) = M[e^{i(\mathbf{t}, \xi)}]$ is the c.f. of the random vector ξ. It is known that the function $\chi_{\xi}(\mathbf{t})$ possesses the following properties:
   a) $\chi_{\xi}(0) = 1$;
   b) $\chi_{\xi}(t)$ is continuous in t;
   c) for an arbitrary collection of vectors $\mathbf{t}^1, ..., \mathbf{t}^k$ and an arbitrary collection of complex numbers $\alpha_1, ..., \alpha_k$ the relation

$$\sum_{n, m=1}^{k} \chi_{\xi}(\mathbf{t}^n - \mathbf{t}^m) \alpha_n \bar{\alpha}_m \geqslant 0$$

is satisfied. Prove that the function

$$\chi(\mathbf{t}) = e^{-(\mathbf{t}, \mathbf{t})/2}$$

satisfies conditions a), b), c).

407. Let ξ_1, ξ_2, ξ_3 be random variables with $M[\xi_i] = 0$ and matrix of second moments $V = \|v_{ik}\|$ and let $1 - p = P\{|\xi_1| < \alpha_1; |\xi_2| < \alpha_2; |\xi_3| < \alpha_3\}$.

Prove that in the case when the ξ_i are not correlated, the estimate

$$p \leqslant \min \left\{ 1, \sum \frac{v_{ii}}{\alpha_i^2} \right\}$$

cannot be strengthened.

408. Let ξ_1 and p be defined as in the preceding problem. Prove that $p \leqslant tr(VB^{-1})$, where B is an arbitrary positive definite matrix with diagonal elements $\leqslant \alpha_j$.

Hint. Use the fact that if in a positive definite quadratic form $S(\mathbf{x}) = \sum a_{ik} x_i x_k$ the diagonal elements of the inverse matrix are $\leqslant 1$, then, for $\mathbf{x} = (x_1, x_2, x_3)$, such that max $(|x_i|) \geqslant 1$, $S(\mathbf{x}) \geqslant 1$.

409. Let there be given n random variables $\xi_1, \ldots, \xi_n$ and let $\rho_{i,j}$ be the correlation coefficient between ξ_i and ξ_j. Prove that the matrix $\|\rho_{i,j}\|_{i,j}^s$ is non-negative definite. In the case $n = 3$, find the possible values for $c = \rho_{12} = \rho_{13} = \rho_{23}$.

410. Prove that for a normal two-dimensional distribution, an arbitrary central moment $\mu_{i,k}$ of even order $i + k = 2n$ equals the coefficient of $t^i u^k$ in the polynomial

$$\frac{i! \, k!}{2^n n!} \left(\mu_{2,0} t^2 + \mu_{1,1} t u + \mu_{0,2} u^2 \right)^n.$$

411. Each of the quantities ξ, η and ζ has mean 0 and dispersion 1. The quantities satisfy the relation $a\xi + b\eta + c\zeta = 0$. Find the matrix of second moments and show that

$$a^4 + b^4 + c^4 \leqslant 2 (a^2 b^2 + a^2 c^2 + b^2 c^2).$$

412. Consider two random variables ξ and η with joint distribution of continuous type. Let

$$f(t, u) = M \left[\exp \left\{ n \left(t\xi + u\eta \right) \right\} \right].$$

Assume also that $\eta > 0$. If the integral

$$g(x) = \frac{1}{2\pi i} \int\limits_{-\infty}^{\infty} \left(\frac{\partial f}{\partial u} \right)_{u = -tx} dt$$

converges uniformly with respect to x, then it represents the probability density of the quantity ξ/η.

413. Suppose the variables $\xi_1, ..., \xi_n$ have a proper normal distribution with means $m_1, ..., m_n$ and matrix of second moments $\Lambda = \|\lambda_{ik}\|$. Prove that the quantity

$$\eta = \sum_{j,k=1}^{n} \frac{\Lambda_{jk}}{|\Lambda|} (\xi_j - m_j)(\xi_k - m_k),$$

where $\Lambda_{j,k}$ is the cofactor of the element $\lambda_{j,k}$, has χ^2 distribution with n degrees of freedom and density

$$k_n(x) = 2^{-n/2} \frac{1}{\Gamma(n/2)} x^{2/n-1} e^{-x/2}$$

for $x > 0$; $k_n(x) = 0$ for $x \leqslant 0$.

See page 141 for the answers on problems 373–413.

8   *Markov chains*

The problems of this chapter correspond basically to §§17–20 of the textbook by B. V. GNEDENKO. Consider the sequence of discrete random variables $\xi_1, ..., \xi_n, ...$. We will say that this sequence forms a *Markov chain* if for an arbitrary finite collection of integers $n_1 < n_2 < \cdots$ $\cdots < n_r < n$ the joint distribution $\xi_{n_1}, \xi_{n_2}, ..., \xi_{n_r}, \xi_n$ is such that the conditional probability of the relation $\xi_n = x$ under the conditions $\xi_{n_1} = = x_1, ..., \xi_{n_r} = x_r$ coincides with the conditional probability $\xi_n = x$ under the condition $\xi_{n_r} = x_r$. Here, $x_1, ..., x_r, x$ are arbitrary numbers for which our conditions have a positive probability. In the case when the transition probabilities

$$p_{jk} = P\{\xi_{m+1} = E_k \mid \xi_m = E_j\}$$

do not depend on m, the Markov chain is called *homogeneous*. The matrix of the transition probabilities of the Markov chain $P = \| p_{ij} \|$ obviously possesses the following properties:

    a) $p_{ik} > 0$;

    b) for all i, $\sum_k p_{ik} = 1$.

Matrices, for which conditions a) and b) are satisfied, are called *stochastic* matrices.

    Sometimes, instead of saying that $\xi_n = E_k$, we say that at the moment of time n the system is in the state E_k. If the number of states in which a system can occur is *finite*, then the Markov chain is called *finite*. We denote

$$p_{i,k}^{(m)} = P\{\xi_n = E_k \mid \xi_{n-m} = E_i\}.$$

If the limit $p_k = \lim_{m \to \infty} p_{ik}^{(m)}$ exists for all k, if this limit does not depend on i, and if $\sum p_k = 1$, then the corresponding Markov chain is said to be *ergodic*. In this connection, the p_i are called *limiting*, or *final*, *probabilities*.

    Let $p_n(j) =$ probability that at time n the Markov chain is in state j.

Then

$$p_n(j) = \sum_{k=1}^{r} p_1(k)\, p_{k,j}^{(n-1)}.$$

If a Markov chain is ergodic then the limits $\lim_{n\to\infty} p_n(j)=p_j$ are also called stationary probabilities.

If a system can go from the state E_i into the state E_j with positive probability, in a finite number of steps, then we say that E_j is accessible from E_i. The state E_i is called *essential* if for every state E_j accessible from E_i, E_i is accessible from E_j. But if at least for one j, E_j is accessible from E_i but E_i is not accessible from E_j, then E_i is an inessential state. A state is called *recurrent* if, starting from it, the system returns to it with probability 1 in a finite number of steps. But if the probability of returning is less than 1, then the state is *transient*. A state is called *periodic* if return to it is possible only after a number of steps which is a multiple of $r>1$. A Markov chain is called *irreducible* if every state of the chain is accessible from any other state. Sometimes Markov chains of the r-th order are also considered. They are defined by the requirement that for an arbitrary collection of integers $n_1 < \cdots < n_s < n,$

$$P\{\xi_{n+r} = x \mid \xi_n = x_n;\ \xi_{n+1} = x_{n+1};\ \ldots;\ \xi_{n+r-1} = x_{n+r-1}\} =$$
$$= P\{\xi_{n+r} = x \mid \xi_{n_1} = x_{n_1};\ \ldots;\ \xi_{n_s} = x_{n_s};\ \xi_n = x_n;\ \ldots;\ \xi_{n+r-1} = x_{n+r-1}\}.$$

where the x, x_i are arbitrary numbers for which the conditions have a positive probability. For $r=1$, we obtain the definition of the usual Markov chain.

The reader is referred to [5], [6] and [7] for additional discussion of Markov chains.

8.1 Definition and examples. Transition probability matrix

414. Let $\xi_k(k=1, 2, \ldots)$ be independent identically distributed integral random variables, where

$$P\{\xi_n = k\} = p_k \quad (k = 0, \pm 1, \pm 2, \ldots),$$

and let $\eta_n = \xi_1 + \xi_n$. Prove that the η_n form a Markov chain. Find the corresponding transition probability matrix.

415. The probabilities of transition after one step in a Markov chain is given by the matrix

$$
P = \begin{bmatrix}
\frac{1}{3} & \frac{1}{3} & \frac{1}{3} & 0 \\
\frac{1}{2} & \frac{1}{2} & 0 & 0 \\
\frac{1}{4} & \frac{1}{4} & 0 & \frac{1}{2} \\
0 & \frac{1}{2} & 0 & \frac{1}{2}
\end{bmatrix}.
$$

a) What is the number of states equal to?

b) How many among them are essential and non-essential?

c) After how many steps from the second state can one go over to the third?

d) Find the transition probabilities after two steps.

416. Let ξ_1, ξ_2, ... be a sequence of integer-valued random variables and let, for an arbitrary $n > 0$ and arbitrary integers $k; j_0; ...; j_n$

$$
P\{\xi_{n+1} = k \mid \xi_0 = j_0; \xi_1 = j_1; ...; \xi_n = j_n\} =
= P\{\xi_{n+1} = k \mid \xi_n = j_n\}.
$$

Prove that in this case, for arbitrary $0 \leqslant n_i \leqslant n$ and arbitrary integers j_i and k

$$
P\{\xi_{n+1} = k \mid \xi_{n_1} = j_1, \xi_{n_2} = j_2, ...\} =
= P\{\xi_{n+1} = k \mid \xi_{n_s} = j_s\}, \quad \text{where} \quad n_s = \max_i \{n_i\}.
$$

417. Prove that for a homogeneous Markov chain with transition probability matrix $P = \|p_{ij}\|$ the following relation holds:

$$
p_{ik}^{(n)} = \sum_j p_{ij} p_{jk}^{(n-1)} \quad (n = 2, 3, ...).
$$

418. Let A be an event depending only on the first $(n-1)$ steps of a Markov chain and B an event depending only on the $(n+1)$, $(n+2)$, ..., $(n+m)$ steps. Prove that for a fixed state at the moment of time n, the events A and B are independent.

419. *Unrestricted random walk.* A point moves on an integral line, passing in one step from the point i to the point $i-1$ with probability p, to the point i with probability q, and to the point $i+1$ with probability

$r(p+q+r=1)$. Find the transition probability matrix: after one step; after two steps.

420. Let ξ_i $(i=1, 2, ...)$ be independent identically distributed random variables which take on the values -1 and $+1$ with probability p and $1-p$ respectively. Set $\zeta_n=\xi_n\xi_{n+1}$. Will the sequence ζ_n be a Markov chain? Does the limit of $P\{\zeta_n=1\}$ exist as $n \to \infty$?

421. Let the ξ_i be defined as in the preceding problem and let

$$\zeta_n = \max_{1 \leqslant k \leqslant n} \sum_{i=1}^{k} \xi_i - \min_{1 \leqslant k \leqslant n} \sum_{i=1}^{k} \xi_i.$$

Will the sequence ζ_n be a Markov chain?

422. There is given a Markov chain with a finite number of states $E_1, ..., E_s$. Let ξ_i be the index of the state in which the system is situated after the i-th step.
Will the following sequences be Markov chains:

a) $\zeta_i = \begin{cases} 1 & \text{if} \quad \xi_i = 1, \\ 0 & \text{if} \quad \xi_i \neq 1 \end{cases}$

b) $\eta_i = \sum_{k=1}^{i} \xi_k$?

Will the second sequence be a chain of order 2?

423. A worker, standing at a control point, inspects one article every minute. Each article independently of the others can turn out to be defective with probability p, $1>p>0$. The worker checks the arriving articles one after the other, spending one minute checking each one. If the article turns out to be defective, then he stops checking the other articles and repairs the defective one. For this he spends 5 minutes. Denote by ζ_n the number of articles accumulated at the worker during n minutes after start of work. Will the sequence ζ_n be a Markov chain?

424. Let ζ_n be defined as in the preceding problem, and let v_n be the time already spent by the worker on checking and repairing the article which at the given moment is serviced by the worker. Will the sequence of two-dimensional random variables (ζ_n, v_n) be a Markov chain?

425. An electron can be situated on one of a countable set of orbits depending on the energy present. Transition from the i-th orbit to the

j-th occurs in one second with probability $C_i \, e^{-\alpha|i-j|}$ $(i, j = 1, 2, ..., \infty)$. Find:

    a) the transition probability after two seconds;

    b) the constants C_i.

426. Suppose the random variables $\xi_0, ..., \xi_n$ form an r-th order Markov chain. Prove that the random vectors $\zeta_n = (\xi_n, \xi_{n+1}, ..., \xi_{n+r-1})$ form a Markov chain of the 1-st order.

427. Prove that if the random variables $\xi_0, \xi_1, ..., \xi_N$ form a Markov chain, then the random variables $\eta_k = \xi_{N-k}$ also form a Markov chain.

428. In studying Markov chains it frequently turns out that one must count the number of transitions from the state E_i to the state E_j. Let the matrix of the initial Markov chain be $P = \|p_{ij}\|_{i,\,j=1}^s$. We introduce a sequence of new random variables $\zeta_1, ..., \zeta_n, ...$ in such a way that ζ_n will take on the value E_{ij} if the initial chain has jumped, at the n-th step, from the state E_i into the state E_j. Prove that the ζ_n also form a Markov chain and find the corresponding probability transition matrix.

429. There are given a sequence ξ_i $(i = 1, 2, ...)$ of independent random variables, each uniformly distributed on the segment $[0, 1]$, a Markov chain with a finite number of states $E_1, ..., E_s$ and transition probability matrix $P = \|p_{ij}\|_{i,\,j=1}^s$. A function $f(\eta, \xi)$ is constructed by defining it in the following way: 1) η takes on the values $E_1, ..., E_s$; 2) if $\eta = E_i$ and $p_{i,1} + \cdots + p_{i,m-1} < \xi \leqslant p_{i,1} + \cdots + p_{i,m}$ then $f(\eta, \xi) = E_m$. Prove that the sequence $\zeta_{n+1} = f(\zeta_n, \xi_{n+1})$ is a Markov chain with transition probability matrix P.

430. Prove: a necessary and sufficient condition for the state E_i in a homogeneous Markov chain to be recurrent is that the series $\sum_{n=1}^{\infty} p_{ii}^{(n)}$ diverge.

Hint. Introduce the probabilities $f_i^{(n)}$ that the first return to the state E_i occurs at the n-th step, and use the formula

$$p_{ii}^{(n)} = \sum_{m=1}^{n} f_i^{(m)} p_{ii}^{(n-m)}.$$

431. Consider the Markov chain with the two states E_1 and E_2, transition probabilities $p_{11} = p_{22} = P$, $p_{12} = p_{21} = q (0 < p < 1, p + q = 1)$ and initial probabilities $P\{\xi_0 = E_1\} = \alpha$, $P\{\xi_0 = E_2\} = 1 - \alpha$. Find $\{p_{ik}^{(n)}\}$, $P_i(n) = P\{\xi_n = E_i\}$ and the corresponding probabilities p_i.

432. N black and N white balls are placed in two urns so that each urn contains N balls. The number of black balls in the first urn determines the state of the system. At each step, one ball is randomly selected from each urn, and these selected balls have their places interchanged. Find p_{ik}. Show that the limiting probability u_k equals the probability of obtaining exactly k black balls if N balls are selected at random from the collection containing N black and N white balls.

433. A Markov chain with states $E_0, E_1, \ldots$ has the transition probabilities

$$
p_{ik} = e^{-\lambda} \sum_{i=0}^{\min(j,\,k)} C_j^i (1 - q)^i q^{j-i} \frac{\lambda^{k-1}}{(k-i)}.
$$

Show that, as $n \to \infty$,

$$
p_{jk}^{(n)} \to e^{-\lambda/q} \left(\frac{\lambda}{q}\right)^k \cdot \frac{1}{k!}.
$$

Such a chain is encountered in statistical mechanics and can be interpreted in the following way. The state of the system is determined by the number of particles in a certain region of space. After every interval of time of unit length, the particle can abandon this region with probability q. Moreover, in this region of space there can appear new particles, and the probability of this is given by the Poisson expression $e^{-\lambda}(\lambda^r/r!)$. The stationary distribution then turns out to be the Poisson distribution with parameter λ/q.

8.2 Classification of states. Ergodicity

434. Are the Markov chains with the following transition probability matrices, after one step, ergodic:

$$
\begin{bmatrix} 0 & 1 \\ 1 & 0 \end{bmatrix}, \quad
\begin{bmatrix} 1 & 0 \\ 1 & 0 \end{bmatrix}, \quad
\begin{bmatrix} 1 & 0 \\ 0 & 1 \end{bmatrix}, \quad
\begin{bmatrix} \frac{1}{2} & \frac{1}{2} \\ 0 & 1 \end{bmatrix}, \quad
\begin{bmatrix} \frac{1}{2} & \frac{1}{2} \\ 1 & 0 \end{bmatrix}?
$$

If they are ergodic, find the limiting distribution.

435. Consider the Markov chain with the states $E_1, \ldots, E_5$ and the transition probability matrix

$$\begin{bmatrix} \frac{1}{2} & \frac{1}{4} & 0 & \frac{1}{4} & 0 \\ 0 & 0 & 1 & 0 & 0 \\ 0 & 0 & 0 & 1 & 0 \\ 0 & 0 & 0 & 0 & 1 \\ 0 & 1 & 0 & 0 & 0 \end{bmatrix}.$$

For $r = 0, 1, 2, 3$, find $a_r = \lim_{n \to \infty} P_{13}^{(4n+r)}$

436. Consider the Markov chain with the following transition probabilities matrix, after one step:

$$P = \begin{bmatrix} p_0 & p_1 \cdots p_{m-1} \\ p_{m-1} & p_0 \cdots p_{m-2} \\ \cdots\cdots\cdots\cdots \\ p_1 & p_2 \cdots p_0 \end{bmatrix},$$

where $0 \leqslant p_i < 1$, $\sum p_i = 1$. Prove that $P\{\xi_n = E_i\} \to 1/m$ as $n \to \infty$.

437. Prove that if the number of states in a Markov chain is $a < \infty$ and if E_k is accessible from E_j, then it is accessible in a or fewer steps.

438. Suppose a chain contains $a < \infty$ states and let E_k be a recurrent state. Prove that there exists a number $q < 1$ for which with $n \geqslant a$ the probability that the time of returning to E_k exceeds n, is less than q^n.

439.* Consider the sequence of random variables $\{\xi_n\}$, defined with the aid of the following recursion formula:

$$\xi_n = \begin{cases} \xi_{n-1} - k + \eta_{n-1} & \text{if } \xi_{n-1} \geqslant k, \\ \xi_{n-1} + \eta_{n-1} & \text{if } \xi_{n-1} < k, \end{cases}$$

where k is a fixed integer and the $\{\xi_n\}$ are independent identically distributed random variables, with

$$P\{\eta_n = l\} = C_m^l p^l (1 - p)^{m-l} \quad (k \neq mp).$$

Prove that the random variables ξ_n form a Markov chain. Show that for $k < mp$ the chain is not ergodic and that for $k > mp$ it is ergodic.
Hint. It is expedient to write out the equation which the $p_j =$

$= \lim_{n \to \infty} P\{\xi_n = j\}$ must satisfy and introduce the generating functions

$$u(z) = \sum_0^\infty p_j z^j \quad \text{and} \quad u_k(z) = \sum_0^{k-1} p_j z^j.$$

440. Random walk with reflecting boundary. Consider the symmetric walk in a bounded region of the plane. The boundary is reflecting in the sense that every time when, in the unrestricted random walk, a particle would leave the region it is forced to return to its previous position. Show that if every point of the region is accessible from every other point then there exists a stationary distribution and that $q_k = 1/a$, where a is a constant in the region.

441. Show that the state E_j of a finite chain is transient if, and only if, there exists an E_k such that E_k is accessible from E_j and E_j is not accessible from E_k. (As the following problem shows, this is not true for an infinite chain.)

442. Assume that in an infinite chain only the transitions $E_j \to E_{j+1}$ and $E_j \to E_0$ are possible and that their probabilities equal $1 - p_j$ and p_j. Show that all states are transient or that all states are recurrent depending on whether the series $\sum p_j$ converges or diverges.

443. Show that an irreducible chain for which at least one diagonal element p_{jj} is positive cannot be periodic.

444. Denote by M_{ij} the mathematical expectation z of the number of steps necessary for the transition from E_i to E_j. Prove that within the bounds of one class of essential states either all the M_{ii} are infinite or all the M_{ii} are finite. The classes in which all the M_{ii} are finite are called *positive* and the classes in which all $M_{ii} = \infty$ are called *null*. Prove that in null classes some M_{ij} can be finite.

445. Prove that a finite irreducible Markov chain is non-periodic if, and only if, there exists a number n such that $p_{ik}^{(n)} > 0$ for all i and k.

446. *Ergodic theorem for means.* For an arbitrary chain, we define the numbers $A_{ik}^{(n)}$ by the equations

$$A_{ik}^{(n)} = \frac{1}{n} \sum_{i=1}^n p_{ik}^{(n)}.$$

Prove that if E_i and E_k belong to the same class of essential states, then $A_{ik}^{(n)}$ tend, as $n \to \infty$, to a limit which does not depend on i. If E_k is irreducible, then $A_{ik}^{(n)} \to 0$ for all i.

447.* Prove that if a characteristic value λ of the finite stochastic matrix is such that $|\lambda| = 1$, then $\lambda = \sqrt[n]{1}$, where n is a natural number.

448.* Prove that if the stochastic matrix has two characteristic values, each in absolute value equal to unity, then the corresponding Markov chain is non-ergodic.

8.3 The distribution of random variables defined on a Markov chain

449.* Let $\eta_n^{(k)} = 1$ if for $t = n$ the chain occurs in E_k and $\eta_n^{(k)} = 0$ otherwise. Set

$$\zeta_n^{(k)} = \sum_{i=1}^{n} \eta_i^{(k)}.$$

Prove that

$$\frac{\zeta_n^{(k)}}{n} \overset{p}{\to} p_k.$$

Hint. Estimate the dispersion $\zeta_n^{(k)}$ and use Chebyshev's inequality.

450.* Denote by $v^{(k)}$ the moment of the first occurrence in E_k under the condition $\xi_0 = k$. Prove that:

a) $P\{v^{(k)} > n\} \leqslant (1 - d)^n$;

b) $M[v^{(k)}] = \dfrac{1}{p_k}.$

Hint. Use the relation $v_k(z) = 1 - 1/u_k(z)$, where $v_k(z)$ is the generating function of $v^{(k)}$, and

$$u_k(z) = 1 + \sum_{n=1}^{\infty} p_{kk}^{(n)} z^n.$$

451. Let $\mu_n^{(k)} = v_0^{(k)} + \cdots + v_n^{(k)}$, where $v_n^{(k)}$ is the time from the n-th occurrence in E_k to the $(n+1)$-st occurrence in E_k inclusive.

Prove that

$$\lim_{n \to \infty} P\left\{ \frac{\mu_n^{(k)} - \dfrac{n}{p_k}}{\sigma\sqrt{n}} < x \right\} = \frac{1}{\sqrt{2\pi}} \int_{-\infty}^{x} e^{-u^2/2}\, du,$$

where $\sigma^2 = D[v_n^{(k)}]$.

452.* We denote by $\zeta_n^{(k)}$ the number of realizations of E_k in the first n steps. Prove that with a suitable normalization $\zeta_n^{(k)}$ converges to the normal law as $n \to \infty$.

Hint. Use the relation

$$P\{\zeta_n^{(k)} < r\} = P\{\mu_r^k > n\},$$

where $\mu_r^{(k)}$ is defined in the preceding problem.

453.* Suppose the transition probability matrix has the form

$$\begin{bmatrix} 1 - \mu & \mu \\ \lambda & 1 - \lambda \end{bmatrix}$$

and let $\xi_n = a_1$ if at the moment of time $t = n$, the chain is in the first state and $\xi_n = a_2 \neq a_1$ if in the second state. Let $\zeta_n = \xi_0 + \cdots + \xi_n$. Prove that

$$\lim_{n \to \infty} P\left\{ \frac{\zeta_n - \left(a_2 + (a_1 - a_2)\dfrac{\lambda}{\lambda + \mu}\right)n}{|a_1 - a_2|\sqrt{\dfrac{n\lambda\mu(2 - \lambda - \mu)}{(\lambda + \mu)^3}}} < x \right\} = \frac{1}{\sqrt{2\pi}} \int_{-\infty}^{x} e^{-u^2/2}\, du.$$

See page 142 for the answers on problems 415–453.

Appendix

Appendix

Table 1   Normal distribution function $\Phi(t)=(1/\sqrt{2\pi})\int_{-\infty}^{t} e^{-\frac{1}{2}x^2}\, dx$

t	0	1	2	3	4
− 0.0	0.5000	0.4960	0.4920	0.4880	0.4840
− 0.1	.4602	.4562	.4522	.4483	.4443
− 0.2	.4207	.4168	.4129	.4090	.4052
− 0.3	.3821	.3783	.3745	.3707	.3669
− 0.4	.3446	.3409	.3372	.3336	.3300
− 0.5	.3085	.3050	.3015	.2981	.2946
− 0.6	.2743	.2709	.2676	.2643	.2611
− 0.7	.2420	.2389	.2358	.2327	.2297
− 0.8	.2119	.2090	.2061	.2033	.2005
− 0.9	.1841	.1814	.1788	.1762	.1736
− 1.0	.1587	.1562	.1539	.1515	.1492
− 1.1	.1357	.1335	.1314	.1292	.1271
− 1.2	.1151	.1131	.1112	.1093	.1075
− 1.3	.0968	.0951	.0934	.0918	.0901
− 1.4	.0808	.0793	.0778	.0764	.0749
− 1.5	.0668	.0655	.0643	.0630	.0618
− 1.6	.0548	.0537	.0526	.0516	.0505
− 1.7	.0446	.0436	.0427	.0418	.0409
− 1.8	.0359	.0351	.0344	.0336	.0329
− 1.9	.0288	.0281	.0274	.0268	.0262
− 2.0	.0228	.0222	.0217	.0212	.0207
− 2.1	.0179	.0174	.0170	.0166	.0162
− 2.2	.0139	.0136	.0132	.0129	.0125
− 2.3	.0107	.0104	.0102	.0099	.0096
− 2.4	.0082	.0080	.0078	.0075	.0073
− 2.5	.0062	.0060	.0059	.0057	.0055
− 2.6	.0047	.0045	.0044	.0043	.0041
− 2.7	.0035	.0034	.0033	.0032	.0031
− 2.8	.0026	.0025	.0024	.0023	.0023
− 2.9	.0019	.0018	.0018	.0017	.0016

$t =$ − 3.0		− 3.1	− 3.2	− 3.3	− 3.4
$(t) =$ 0.0013		0.0010	0.0007	0.0005	0.0003

5	6	7	8	9
0.4801	0.4761	0.4721	0.4681	0.4641
.4404	.4364	.4325	.4286	.4247
.4013	.3974	.3936	.3897	.3859
.3632	.3594	.3557	.3520	.3483
.3264	.3228	.3192	.3156	.3121
.2912	.2877	.2843	.2810	.2776
.2578	.2546	.2514	.2483	.2451
.2266	.2236	.2206	.2177	.2148
.1977	.1949	.1922	.1894	.1867
.1711	.1685	.1660	.1635	.1611
.1469	.1446	.1423	.1401	.1379
.1251	.1230	.1210	.1190	.1170
.1056	.1038	.1020	.1003	.0985
.0885	.0869	.0853	.0838	.0823
.0735	.0721	.0708	.0694	.0681
.0606	.0594	.0582	.0571	.0559
.0495	.0485	.0475	.0465	.0455
.0401	.0392	.0384	.0375	.0367
.0322	.0314	.0307	.0301	.0294
.0256	.0250	.0244	.0239	.0233
.0202	.0197	.0192	.0188	.0183
.0158	.0154	.0150	.0146	.0143
.0122	.0119	.0116	.0113	.0110
.0094	.0091	.0089	.0087	.0084
.0071	.0069	.0068	.0066	.0064
.0054	.0052	.0051	.0049	.0048
.0040	.0039	.0038	.0037	.0036
.0030	.0029	.0028	.0027	.0026
.0022	.0021	.0021	.0020	.0019
.0016	.0015			
− 3.5	− 3.6	− 3.7	− 3.8	− 3.9
0.0002	0.0002	0.0001	0.0001	0.0000

Table 2　Student's criterion. Confidence boundaries for t with f degrees of freedom

f ↓	Two-sided boundaries			
	5%	2%	1%	0.1%
1	12.710	31.820	63.660	636.600
2	4.303	6.965	9.925	31.600
3	3.182	4.541	5.841	12.920
4	2.776	3.747	4.604	8.610
5	2.571	3.365	4.032	6.869
6	2.447	3.143	3.707	5.959
7	2.365	2.998	3.499	5.408
8	2.306	2.896	3.355	5.041
9	2.262	2.821	3.250	4.781
10	2.228	2.764	3.169	4.587
11	2.201	2.718	3.106	4.437
12	2.179	2.681	3.055	4.318
13	2.160	2.650	3.012	4.221
14	2.145	2.624	2.977	4.140
15	2.131	2.602	2.947	4.073
16	2.120	2.583	2.921	4.015
17	2.110	2.567	2.898	3.965
18	2.101	2.552	2.878	3.922
19	2.093	2.539	2.861	3.883
20	2.086	2.528	2.845	3.850
21	2.080	2.518	2.831	3.819
22	2.074	2.508	2.819	3.792
23	2.069	2.500	2.807	3.767
24	2.064	2.492	2.797	3.745
25	2.060	2.485	2.787	3.725
26	2.056	2.479	2.779	3.707
27	2.052	2.473	2.771	3.690
28	2.048	2.467	2.763	3.674
29	2.045	2.462	2.756	3.659
30	2.042	2.457	2.750	3.646
↑ f	2.5%	1%	0.5%	0.05%
	One-sided boundaries			

Table 2 (continued)

f ↓	Two-sided boundaries			
	5%	2%	1%	0.1%
40	2.021	2.423	2.704	3.551
50	2.009	2.403	2.678	3.495
60	2.000	2.390	2.660	3.460
80	1.990	2.374	2.639	3.415
100	1.984	2.365	2.626	3.389
200	1.984	2.365	2.626	3.389
500	1.965	2.334	2.586	3.310
∞	1.960	2.326	2.576	3.291
↑ f	2.5%	1%	0.5%	0.05%
	One-sided boundaries			

Table 3   Poisson distribution. The function $\sum_{k=x}^{\infty} (\lambda^k/k!)\, e^{-\lambda}$

λ					
x	0.1	0.2	0.3	0.4	0.5
0	1.000000	1.000000	1.000000	1.000000	1.000000
1	.095163	.181269	.259182	.329680	.393469
2	.004679	.017523	.036936	.061552	.090204
3	.000155	.001149	.003600	.007926	.014388
4		.000057	.000266	.000776	.001752
5					.000172

λ					
x	0.6	0.7	0.8	0.9	1.0
0	1.000000	1.000000	1.000000	1.000000	1.000000
1	.451188	.503415	.550671	.593430	.632121
2	.121901	.155805	.191208	.227518	.264241
3	.023155	.034142	.047423	.062857	.080301
4	.003358	.005753	.009080	.013459	.018988
5	.000394	.000786	.001411	.002344	.003660
6			.000184	.000343	.000594

Table 3 (continued)

λ					
x	1.2	1.4	1.6	1.8	2.0
0	1.000000	1.000000	1.000000	1.000000	1.000000
1	.698806	.753403	.798103	.834701	.864665
2	.337373	.408167	.475069	.537163	.593994
3	.120513	.166502	.216642	.269379	.323324
4	.033769	.053725	.078813	.108708	.142877
5	.007746	.014253	.023682	.036407	.052653
6	.001500	.003201	.006040	.010378	.016564
7	.000251	.000622	.001336	.002569	.004534
8			.000260	.000562	.001097
9					.000237

λ					
x	2.2	2.4	2.6	2.8	3.0	4.0
0	1.000000	1.000000	1.000000	1.000000	1.000000	1.000000
1	.889197	.909282	.925726	.939190	.950213	.981684
2	.645430	.691559	.732615	.768922	.800852	.908422
3	.377286	.430291	.481570	.530546	.576810	.761897
4	.180648	.221227	.263998	.308063	.352768	.566530
5	.072496	.085869	.122577	.152324	.184737	.371163
6	.024910	.035673	.049037	.065110	.083918	.214870
7	.007461	.011594	.017170	.024411	.033590	.110674
8	.001978	.003339	.005334	.008131	.011905	.051134
9	.000470	.000862	.001487	.002433	.003803	.021363
10			0.000376	0.000660	0.001102	0.008132
11					.000292	.002840
12						.000915

Table 3 (continued)

λ						
x	5.0	6.0	7.0	8.0	9.0	10.0
0	1.000000	1.000000	1.000000	1.000000	1.000000	1.000000
1	.993262	.997521	.999088	.999665	.999877	.999955
2	.959572	.982649	.922705	.996981	.998766	.999501
3	.875348	.938031	.970364	.986246	.993768	.997231
4	.734974	.848796	.918235	.957620	.978774	.989661
5	.559507	.714943	.827008	.900368	.945036	.970747
6	.384039	.554320	.299292	.808764	.884309	.932914
7	.237817	.392697	.550289	.686626	.793219	.869859
8	.133372	.256020	.401286	.547039	.676103	.779779
9	.068094	.152763	.270909	.407453	.544347	.667180
10	.031828	.083924	.169504	.283376	.412592	.542070
11	.013695	.042621	.098521	.184114	.294012	.416960
12	.005453	.020092	.053350	.111924	.196662	.303224
13	.002019	.008827	.027000	.063797	.124227	.208444
14	.000698	.003628	.012811	.034181	.073851	.135536
15		.001400	.005717	.017257	.041466	.083459
16		.000509	.002407	.008231	.022036	.048740
17			.000958	.003718	.011106	.027042
18				.001594	.005320	.014278
19				.000650	.002426	.007187
20					.001056	.003454
21					.000439	.001588
22						.000700

Appendix

Table 4 Confidence boundaries for χ^2 with f degrees of freedom

f	5%	1%	0.1%	f	5%	1%	0.1%
1	3.84	6.63	10.8	18	28.9	34.8	42.3
2	5.99	9.21	13.8	19	30.1	36.2	43.8
3	7.81	11.3	16.3	20	31.4	37.6	45.3
4	9.49	13.3	18.5	21	32.7	38.9	46.8
5	11.1	15.1	20.5	22	33.9	40.3	48.3
6	12.6	16.8	22.5	23	35.2	41.6	49.7
7	14.1	18.5	24.3	24	36.4	43.0	51.2
8	15.5	20.1	26.1	25	37.7	44.3	52.6
9	16.9	21.7	27.9	30	43.8	50.9	59.7
10	18.3	23.2	29.6	35	49.8	57.3	66.6
11	19.7	24.7	31.3	40	55.8	63.7	73.4
12	21.0	26.2	32.9	45	61.7	70.0	80.1
13	22.4	27.7	34.5	50	67.5	76.2	86.7
14	23.7	29.1	36.1	55	73.3	82.3	93.2
15	25.0	30.6	37.7	60	79.1	88.4	99.6
16	26.3	32.0	39.3	65	84.8	94.4	106.0
17	27.6	33.4	40.8	70	90.5	100.4	112.3

Answers

1   Fundamental concepts

1. a) A young man who does not live in a dormitory and does not smoke is chosen.

b) When all young men live in a dormitory and do not smoke.

c) When the smokers live only in the dormitory.

d) When no young lady smokes and all young men smoke. No, since young ladies can also smoke.

9. a) $B_{4,2} = (A_1 \cap A_2 \cap \bar{A}_3 \cap \bar{A}_4) \cup (A_1 \cap \bar{A}_2 \cap A_3 \cap \bar{A}_4) \cup$
$\cup (A_1 \cap \bar{A}_2 \cap \bar{A}_3 \cap A_4) \cup (\bar{A}_1 \cap A_2 \cap A_3 \cap \bar{A}_4) \cup$
$\cup (\bar{A}_1 \cap A_2 \cap \bar{A}_3 \cap A_4) \cup (\bar{A}_1 \cap \bar{A}_2 \cap A_3 \cap A_4);$

b) if the experiment $\mathfrak{A}$ is repeated indefinitely the event A occurs m times;

c) true.

17. 10%.

21. 113.

23. $P\{A \bigtriangleup B\} = 2r - p - q; \quad P\{A \cap \bar{B}\} = r - q;$
$P\{\bar{A} \cap \bar{B}\} = 1 - r.$

27. The space of elementary outcomes consists of the sequence:
TT, HH, THH, HTT, THTT, HTHH,...

a) $\frac{15}{16}$;

b) $\frac{2}{3}$.

29. $\dfrac{3!2!2!}{10!}.$

31. 0.096.

33. a) $\dfrac{C_{51}^5}{C_{52}^6}$;

b) $\dfrac{C_4^1(C_{13}^1)^3 C_{13}^3 + C_4^2[C_{13}^1 \cdot C_{13}^2]^2}{C_{52}^6}$;

c) to find n, one must solve the equation $1 - \dfrac{(C_4^1)^n C_{13}^n}{C_{52}^n} > \frac{1}{2}$.

35. $\dfrac{2(k-1)(n-k)}{n(n-1)}$.

37. $1 - \dfrac{C_{n-m}^k}{C_n^k}$.

39. $\dfrac{C_M^m C_{N-M}^{n-m}}{C_N^n}$.

41. a) $\dfrac{C_n^{2r} 2^{2r}}{C_{2n}^{2r}}$;

b) $\dfrac{n C_{n-1}^{2r-2} 2^{2r-2}}{C_{2n}^{2r}}$;

c) $\dfrac{C_n^2 C_{n-2}^{2r-4} 2^{2r-4}}{C_{2n}^{2r}}$.

43. Consider random walk on a two-dimensional integral lattice of points $\xi(l)$ defined in the following way: $\xi(0)=(0, 0)$, $\xi(l)=(M(l), N(l))$ for $0 < l \leqslant m+n$. Clearly, all trajectories of the point ξ emanating from the point $(0, 0)$ to the point (m, n) are equally probable. The total number of such trajectories is C_{m+n}^m. Denote by z the number of trajectories going from $(0, 0)$ to (m, n) and having at least one point in common with the line $x=y$, different from $(0, 0)$. Then

$$P = 1 - \frac{z}{C_{n+m}^m}.$$

In order to find z, we note that between the set of trajectories going from $(1, 0)$ to (m, n) and the set of trajectories going from $(0, 1)$ to (m, n) and having at least one point in common with the line $x=y$ one can establish a one-to-one correspondence. To this end, it suffices to assign to each trajectory emanating from $(1, 0)$ the trajectory emanating from $(0, 1)$, which is obtained from the first by a mirror reflection with respect to the line $x = y$ of the segment of the first trajectory from the point $(1, 0)$ to the first point in common with the line $x = y$. Now it is easy to find z:

$$z = 2C_{n+m-1}^n.$$

Answer: $P = \dfrac{n - m}{n + m}.$

45. a) $\dfrac{1}{1 \cdot 3 \cdot 5 \ldots (2n - 1)} = \dfrac{2^n \cdot n!}{(2n)!}$;

b) $\dfrac{n!}{1 \cdot 3 \cdot 5 \ldots (2n - 1)} = \dfrac{n!}{(2n - 1)!!} = \dfrac{2^n}{C_{2n}^n}.$

47. $P_n = \dfrac{C_4^n \displaystyle\sum_{k=0}^{4-n} C_{4-n}^k \cdot 2^k \cdot C_{44}^{13-k-2n}}{C_{52}^{13}}.$

49., 51. The solution of these problems leans on the following lemma. *In all, there exist C_{n+r-1}^r different ways to arrange r non-distinct objects in n cells.*

Proof. Think of the cells as the intervals between $(n + 1)$ marks and for the objects take the letter A. The symbol $|AA| \, |A| \, |AAAA|$ will mean that there are 6 cells and 7 objects altogether, where there are 2 objects in the first cell, 1 object in the third cell, 4 in the sixth cell; cells 2, 4, 5 are empty. The distribution of objects in the cells is fixed as soon as the places on which the letters occur are indicated. But the letters can occur at $(n + r - 1)$ places, and therefore the number of different arrangements coincides with

$$C_{n+r-1}^r.$$

Answers:

49. C_{n+r-1}^r;

51. $1 : C_{n+r-1}^r.$

55. $n^{-k} C_n^r [C_r^0 r^k - C_r^1 (r - 1)^k + C_r^2 (r - 2)^k - \cdots - (-1)^{r-1} C_r^{r-1} 1^k].$

127

Answers

57. a) $\left(\dfrac{r}{R}\right)^2$;

b) $\dfrac{2\alpha}{\pi}$.

59. a) $1 - (1 - z)^2$;
b) $z(1 - 1nz)$;
c) $1 - (1 - z)^2$;
d) z^2 ;
e) for $z < \tfrac{1}{2}$, $2z^2$; for $z > \tfrac{1}{2}$, $1 - 2(1 - z)^2$.

61. a) $\dfrac{1}{a^2}(a - 2r)^2$;

b) $1 - 4\left(\dfrac{r}{a}\right)^2$.

63. for $0 \leqslant x \leqslant k$ and $\arctan \dfrac{k}{l} \leqslant y \leqslant \arctan \dfrac{k}{l}$;

$$P\{h < x, \alpha < y\} = \dfrac{x}{kl}(2l - x \cot y).$$

for $0 \leqslant x \leqslant k$ and $0 \leqslant y < \arctan \dfrac{x}{l}$;

$$P\{h < x, \alpha < y\} = \dfrac{l \tan y}{k}.$$

2   Application of the basic formulas

71. $\frac{20}{25} \cdot \frac{19}{24} \cdot \frac{18}{23}$.

73. $P\{B \mid A\} = 0.39$; $P\{B \mid \bar{A}\} = 0.10$.

75. $1 - (\tfrac{3}{4})^3$.

77. $\frac{20}{21}$.

79. $\frac{11}{17}$.

81. $\dfrac{C_5^2}{C_6^3}$.

83. Dependent.

85. For $p = \frac{1}{2}$.

87. $\dfrac{p - \varepsilon}{1 - \varepsilon} \leqslant P\{A \mid B\} \leqslant \dfrac{p}{1 - \varepsilon}$.

89. Let ξ, η, ζ be defined as in problem 85 and let $p = \frac{1}{2}$; then the events $A = \{\xi = 0\}$, $B = \{\eta = 0\}$, $C = \{\zeta = 0\}$ are pairwise independent. On the other hand, $A \cap B \subset C$ and hence A, B, C are simultaneously dependent.

91. $x = \frac{1}{2}$.

93. $P_k = 2^{-k}$. 2 times.

95. $2^{-5}(C_5^0 + C_5^1)$.

97. a) $P_{3,4} = C_4^3 \cdot \left(\frac{1}{2}\right)^3 \cdot \frac{1}{2} = \frac{1}{4}$; $P_{5,8} = C_8^5 \left(\frac{1}{2}\right)^5 \cdot \left(\frac{1}{2}\right)^3 = \frac{7}{32}$;

$P_{3,4} > P_{5,8}$;

b) $P_{3,4} + P_{4,4} = \frac{5}{16}$;

$P_{5,8} + P_{6,8} + P_{7,8} + P_{8,8} = (C_8^3 + C_8^2 + C_8^1 + C_8^0)\,2^{-8} =$

$= \frac{93}{256} > \frac{5}{16}$;

c) $\displaystyle\sum_{k=0}^{n} P_{k,2n} = (1 + C_{2n}^1 + \cdots + C_{2n}^n)\,2^{-2n} >$

$> (1 + C_{2n}^1 + \cdots + C_{2n}^{n-1})\,2^{-2n} =$

$= (C_{2n}^{n+1} + \cdots + C_{2n}^{2n})\,2^{-2n} = \displaystyle\sum_{k=n+1}^{2n} P_{k,2n}$;

d) equally probable.

101. $C_{10}^3\,(0.3)^3 \cdot (0.7)^7$; $\displaystyle\sum_{k=0}^{3} C_{10}^k\,(0.3)^k \cdot (0.7)^{10-k}$.

103. The second player must call zero with probability 1. The probability that k will be guessed between two successive failures equals 2^{-k-1}.

105. a) $\dfrac{2}{\pi}$;

b) $\dfrac{10! \cdot 4}{4!\,3!\,1!\,1!\,1!} \left(\dfrac{2}{\pi}\right)^{4} \left(1 - \dfrac{2}{\pi}\right)^{6} \cdot 4^{-6}$.

107. $P_A = (0.8 \cdot 0.4 \cdot 0.4) + (0.8 \cdot 0.6 \cdot 0.4) + (0.8 \cdot 0.4 \cdot 0.6) +$
$+ (0.2 \cdot 0.4 \cdot 0.4) = 0.544 > \tfrac{1}{2}$.

109. $P\{A_k\} = (1 - q)^k$; $(1 - q)^l q$.

113. $\tfrac{1}{2}$.

115. a) $\tfrac{1}{4}$;

b) $3 \ln 2 - 2 = 0.082 \ldots$.

117. $e^{-\lambda t}$.

119. For $0 \leqslant x \leqslant k \dfrac{1}{\sqrt{3}} \; P\left\{h < x \mid \alpha \leqslant \dfrac{\pi}{6}\right\} = 1 - \left(1 - \dfrac{x\sqrt{3}}{k}\right)^{2}$.

121. a) $\left(1 - \dfrac{r^3}{R^3}\right)^{N}$;

b) $e^{-\frac{4}{3}\pi r^3 \lambda}$.

123. 0.52.

125. Immaterial.

127. $1 - (1 - p(1 - q))^{n}$.

135. a) $1 - \displaystyle\prod_{i=1}^{n} (1 - p_i)$;

b) $\displaystyle\prod_{i=1}^{n} (1 - p_i)$;

c) $\displaystyle\sum_{i=1}^{n} p_i \prod_{j \neq i} (1 - p_j)$.

137. One must use the formula in Problem 136(a):

$$P_n = 1 - \frac{1}{2!} + \frac{1}{3!} - \cdots \pm \frac{1}{n!};$$

$$\lim_{n \to \infty} P_n = 1 - \frac{1}{e}.$$

139. The number of terms in the development of a determinant of the n-th order is equal in general to $n!$. The number of terms containing a given element equals $(n-1)!$. The number of terms containing two given elements equals $(n-2)!$, etc. Hence

$$P_n = 1 - \frac{1}{1!} + \frac{1}{2!} - \frac{1}{3!} + \cdots + (-1)^n \frac{1}{n!},$$

$$\lim_{n \to \infty} P_n = e^{-1}.$$

141. $e^{-\lambda t}$.

143. $p_r'(t) = - brp_r(t) + a(r-1)p_{r-1}(t)$.

145. $\dfrac{a}{a+b}$.

147. $\beta = 21$.

3   # Random variables and their properties

149. Let ξ be the length of the game; then $P\{\xi=k\}=2^{-k}$ and $M[\xi]=2$.

151. 10.

153. $3.5n$; $\frac{1}{6} \cdot 17.5n$;   0.

155. ≈ 8.50; $\frac{1}{24}$.

157. λp.

Answers

159. Let S be the area of the circle and d its diameter. For

$$\frac{\pi a^2}{4} \leqslant x \leqslant \frac{\pi b^2}{4}; \quad P\{S < x\} = \frac{1}{b - a}\left(\sqrt{\frac{4x}{\pi}} - a\right).$$

$$M[S] = M\left[\frac{\pi}{4} d^2\right] = \frac{\pi}{4} M[d^2] = \frac{\pi}{12}(b^2 + ab + a^2).$$

$$D[S] = D\left[\frac{\pi}{4} d^2\right] = \frac{\pi^2}{16} D[d^2] = \frac{\pi^2}{16}(M[d^4] - (M[d]^2) =$$

$$= \frac{\pi^2}{720}(b - a)^2(4b^2 + 7ab + 4a^2).$$

161. $\dfrac{1}{0.003}(1 - e^{-1.095}) \approx 222$ (days).

167. $M[\xi\eta] = 1; \quad D[\xi\eta] = \frac{4}{9}.$

169. $M[\zeta] = \displaystyle\sum_{i=1}^{n} M[\eta_i] = 2npq;$

$$D[\zeta] = \sum_{i=1}^{n} D[\eta_i] + 2\sum_{i=1}^{n-1} \text{cov}(\eta_i, \eta_{i+1}) = n2pq(1 - 2pq) +$$

$$+ (n - 1) 2pq(p - q)^2.$$

173. a) 1;

b) $\dfrac{\beta^{\alpha+1}}{\Gamma(\alpha + 1)}$;

c) $\dfrac{1}{\pi}$.

183. a) Let $\zeta_1 = \alpha\xi + \beta\eta; \quad \zeta_2 = \alpha\xi - \beta\eta.$

$$D[\zeta_1] = D[\zeta_2] = (\alpha^2 + \beta^2)\sigma^2; \quad \rho(\zeta_1, \zeta_2) = \alpha^2 - \beta^2;$$

$$p_{\xi_1, \xi_2}(x, y) = \frac{1}{2\pi\sigma^2(\alpha^2 + \beta^2)(1 - \rho^2)} \times$$

$$\times \exp\left\{-\frac{(x - (a + \beta)a)^2 - 2\rho(x - (\alpha + \beta)a) \times}{2\sigma^2(\alpha^2 + \beta^2)(1 - \rho^2)}\right\}.$$

132

187. For $x < 0$ this follows from the inequalities

$$-x = \int_{-\infty}^{-\infty} (y - x)\, dF(y) \leqslant \int_{x}^{\infty} (y - x)\, dF(y)$$

$$x^2 \leqslant \left(\int_{x}^{\infty} (y - x)\, dF(y) \right)^2 \leqslant \int_{x}^{\infty} dF(y) \cdot \int_{x}^{\infty} (y - x)^2\, dF(y) \leqslant$$

$$\leqslant (1 - F(x))\,(\sigma^2 + x^2).$$

For $x > 0$ the proof is analogous.

191. $\dfrac{1}{2h} \displaystyle\int_{-h}^{h} F(x - y)\, dy$.

197. a) $C = \dfrac{1}{\pi}$;

b) the same as for ξ;

c) $\tfrac{1}{4}$.

199. a) For $x > 0$, $p_\zeta(x) = \dfrac{1}{(1 + x)^2}$;

b) for $0 < x < 1$, $p_\zeta(x) = \tfrac{1}{2}$;

   for $x > 1$, $\quad p_\zeta(x) = \dfrac{1}{2x^2}$;

c) $p_\zeta(x) = \dfrac{1}{\pi(1 + x^2)}$.

203. a) $\Psi(x)$ is the function inverse to

$$\Phi(x) = \frac{1}{\sqrt{2\pi}} \int_{\infty}^{x} \exp\left\{ -\frac{u^2}{2} \right\} du.$$

207. $P\{v(x) = k\} = C_n^k F^k(x)\,(1 - F(x))^{n-k}$.

209. k weighings, where k is defined by the condition $3^{k-1} < n \leqslant 3^k$; $\log_2 3$ double units.

213. $H = -\dfrac{p \log p + (1 - p) \log (1 - p)}{p}$; as p decreases from 1 to 0

the entropy increases from 0 to ∞.

219. $I_{YZ}(X) = - p \log p - (1 - p) \log (1 - p) +$

$$+ p\varDelta^2 \log \frac{p\varDelta^2}{p\varDelta^2 + (1 - p)(1 - \delta)^2} +$$

$$+ (1 - p)(1 - \delta)^2 \log \frac{(1 - p)(1 - \delta)^2}{p\varDelta^2 + (1 - p)(1 - \delta)^2} +$$

$$+ 2p\varDelta (1 - \varDelta) \log \frac{p\varDelta (1 - \varDelta)}{p\varDelta (1 - \varDelta) + (1 - p)(1 - \delta)\delta} +$$

$$+ 2(1 - p)(1 - \delta)\delta \log \frac{(1 - p)(1 - \delta)\delta}{p\varDelta (1 - \varDelta) + (1 - p)\delta(1 - \delta)} +$$

$$+ p(1 - \varDelta)^2 \log \frac{p(1 - \varDelta)^2}{p(1 - \varDelta)^2 + (1 - p)\delta^2} +$$

$$+ (1 - p)\delta^2 \log \frac{(1 - p)\delta^2}{p(1 - \varDelta)^2 + (1 - p)\delta^2};$$

$I_{YZ}(X) = 0.2234$ double units.

4 Basic limit theorems

223. Let S be the number of boys among the newborn. It is necessary to find the probability that $S \leqslant 5000$. Since $M[S] = 5150$ and $D[S] \approx 2500$, the probability sought equals $\varPhi(-3) = 0.0013$.

225. $\dfrac{1,359,671}{1,285,086 + 1,359,671} = 0.5141$;

$$\sqrt{\frac{p(1 - p)}{n}} \approx \sqrt{\frac{1}{4n}} = 0.0003.$$

According to the criterion, 0.5 is incompatible; 0.515 is compatible.

227. Apply the Poisson approximation. 0.93803; 0.99983; 0.16062.

229. Apply the Poisson distribution. 0.00016.

231. a) $\dfrac{8,506}{34,153} \approx 0.249$; $\quad 0.249 + \alpha \sqrt{\dfrac{0.249 \cdot 0.751}{34,153}} = 0.245, \alpha = -1.71$;

$0.249 + \beta \sqrt{\dfrac{0.249 \cdot 0.751}{34,153}} = 0.255, \beta \approx 2.56$;

$P(0.245 < p < 0.255) \approx \Phi(2.56) - \Phi(-1.71) \approx 0.951$;

b) $1 - \Phi(0.401) + \Phi(-0.401) \approx 0.688$;

c) $\Phi\left(\dfrac{0.01\sqrt{n}}{\sqrt{\frac{1}{4} \cdot \frac{3}{4}}}\right) \geqslant 0.995$. It is sufficient to take $n \geqslant 12,500$.

233. $\frac{1}{2}$; the probability of obtaining a worse (in absolute value) coincidence of the number of guesses with 50 is about 0.05. The result can be ascribed a purely random coincidence.

235. $e^{-x} \leqslant 0.01$; $\quad x \geqslant 5$.

237. $e^{-x}\left(1 + x + \dfrac{x^2}{2} + \dfrac{x^3}{3}\right) \leqslant 0.01$; $\quad n \cong x \cdot 10,000 \approx 107,000$.

241. Denote by $\xi_i (i=1, 2, 3)$ the number of parts in disrepair in the i-th group. From the conditions of the problem it follows that the ξ_i have approximately a Poisson distribution with parameters $\lambda_1 = 0.3$; $\lambda_2 = 0.4$; $\lambda_3 = 0.7$. The number of parts in disrepair in the machine $\xi = \xi_1 + \xi_2 + \xi_3$ also has a Poisson distribution with parameter $\lambda = 1.4$,

$$P\{\xi \geqslant 2\} = 1 - e^{-\lambda}(1 + \lambda) = 0.408.$$

243. If the player wrote down k numbers, then the probability p_k that all the numbers he wrote down will occur among the five coming out in the drawing equals

$$p_k = \dfrac{C_{90-k}^{5-k}}{C_{90}^{5}};$$

$$p_1 = \dfrac{1}{18}, \; p_2 = \dfrac{2}{801}, \; p_3 = \dfrac{1}{11,748}, \; p_4 = \dfrac{1}{511,038}, \; p_5 = \dfrac{1}{43,949,268}.$$

Denote the mean value of the player's win, who wrote down k numbers

Answers

by E_k; if the wager equals a rubles, then

$$E_1 = 15a \cdot \tfrac{1}{18} - a \cdot 1 = -\tfrac{1}{6}a, \quad E_2 = -\tfrac{29}{89}a \approx -\tfrac{1}{3}a, \quad \text{etc}.$$

Since all the E_k are <0, it is obvious that the lottery is a game which is unfavorable upon writing down an arbitrary number of numbers. The probability that the number of winners among those who wrote down three numbers will be greater than 10 equals ≈ 0.24.

261. $\dfrac{|x - 3500|}{\sqrt{1000 \cdot \frac{1}{6} \cdot 17.5}} < 1.96; \quad |x - 3500| < 106.$

263. $\pm 0.866 \cdot 10^{-m+2}$

265. $\approx 0.24; \quad \approx 3840.$

267. a) No;

  b) LLN yes; CLT no;

  c) LLN no; CLT yes.

271. $0.08; \quad 0.0013; \quad 0.85.$

273. $M[I_N] = I; \quad D[I_N] \leqslant \dfrac{4a^2}{N}; \quad$ normal with parameters $(0, \sigma)$,

where

$$\sigma^2 = \iiint\limits_V (f(\mathbf{x}) - I)^2 \, dV.$$

275. The quantities $\dfrac{\chi_n^2 - n}{\sqrt{2n}}$ and τ_n are asymptotically $N(0, 1)$.

277. $N(0, 1)$.

279. Let η_i be the number of people who have passed by the seller during the time from the sale of the $(i-1)$-st paper to the time of sale of the i-th paper; the $\eta_i (i=1, 2, ..., n)$ are independent and identically distributed, $\xi = \sum \eta_i$. Apply the CLT. $M[\xi] = 300, D[\xi] = 100, D[\eta] = 900$. The quantity $(\xi - 300)/30$ is asymptotically $N(0, 1)$.

| 5 | **Characteristic and generating functions** |

281. a) $\cos t = \frac{1}{2} e^{it(-1)} + \frac{1}{2} e^{it(1)}$.

Consequently,

$$F(x) = \begin{cases} 0 & x \leqslant -1 \\ \frac{1}{2} & -1 < x \leqslant 1; \\ 1 & x > 1 \end{cases}$$

b) $\cos^2 t = \frac{1}{4} e^{it(-2)} + \frac{1}{2} + \frac{1}{4} e^{it2}_{}$;

$$F(x) = \begin{cases} 0 & x \leqslant -2 \\ \frac{1}{4} & -2 < x \leqslant 0 \\ \frac{3}{4} & 0 < x \leqslant 2 \\ 1 & 2 < x \end{cases}$$

c) $\sum a_k \cos kt = \frac{1}{2} \sum a_k e^{-ikt} + \frac{1}{2} \sum a_k e^{ikt}$.

Discrete distribution with jumps $\frac{1}{2}a_k$ at the points $\pm k$.

283. a) $\frac{1}{4}(1 + z)^2 = \frac{1}{4} + \frac{1}{2}z + \frac{1}{4}z^2$.

Discrete distribution, with jumps at the points 0; 1; 2; respectively equal to $\frac{1}{4}$; $\frac{1}{2}$; $\frac{1}{4}$;

b) $\frac{1}{2}(1 - \frac{1}{2}z)^{-1} = \frac{1}{2} + \frac{1}{4}z + \frac{1}{8}z^2 + \cdots$.

Discrete distribution. The value $k \geqslant 0$ is taken on with probability 2^{-k-1}.

c) Poisson distribution with parameter λ;

d) binomial distribution with probability of success $p = \frac{2}{3}$.

285. $\phi(z) = \dfrac{p(1-p)z^2}{(1-pz)(1-(1-p)z)}$; $M[U_n] = \dfrac{1}{p(1-p)}$;

$D[U_n] = \dfrac{1 - 3p(1-p)}{p^2(1-p)^2}$.

287. $\phi(z) = \left(\dfrac{1-p}{1-pz}\right)^r$; $P_{r,k} = (1-p)^r p^k C^k_{r+k-1}$.

291. a) Let $p(x)$ be the density. For an arbitrary $\varepsilon > 0$ one can find a step function $\phi(x)$, having a finite number of jumps, such that

$$\int |p(x) - \phi(x)|\, dx < \varepsilon.$$

The integral $\int e^{itx} \phi(x)$ can be written in the form of a finite sum of the form

$$\sum C_u \int_{au}^{bu} e^{-itx}\, dx.$$

Every term of this sum tends to zero as $t \to \infty$. Consequently, there exists a T such that, for $|t| > T$,

$$\left| \int_x e^{itx} p(x)\, dx \right| < 2\varepsilon.$$

293. a) $\dfrac{1}{2T} \displaystyle\int_{-T}^{T} f(t)\, e^{-itx}\, dt = \dfrac{1}{2T} \int_{-T}^{T} dt \int_{-\infty}^{\infty} e^{it(y-x)}\, dF(y) =$

$$= \int_{-\infty}^{\infty} \frac{\sin Ty}{Ty}\, d_y F(y + x) = \int_{-h}^{h} + \int_{-\infty}^{h} + \int_{h}^{\infty}.$$

For an arbitrary h, the second and third integrals tend to zero as $T \to \infty$. If h is chosen sufficiently small, then as $T \to \infty$ the limit of the first integral will differ by an arbitrarily small amount from the jump of the function F at the point x.

b) For the proof, it suffices to note that the limit of the integral on the left equals, according to a), above, the jump at zero of the distribution function $F^*(x) = F(x)^*[1 - F(-x+0)]$ and to calculate the magnitude of the jump.

295. Cases a), d), e) will be stable.

297. $\left(\dfrac{\alpha}{1+\alpha}\right)^{\lambda}\left(1 - \dfrac{e^{it}}{1+\alpha}\right)^{-\lambda}$; $\dfrac{\lambda}{\alpha}$; $\dfrac{\lambda}{\alpha}\left(1 + \dfrac{\lambda}{\alpha}\right)$.

305. LLN is applicable for $\alpha < \frac{1}{2}$. CLT is applicable for all values of α.

323. $(1 + yx)(1 + yx^2) \cdots (1 + yx^N)$; $\frac{3}{20}$.

138

325. $\dfrac{2^n}{(n+1)!}$.

6   Application of measure theory

333. $A_k = \left(\bigcap_{n=1}^{\infty} \{\omega : \zeta_n \leqslant k\}\right) - \left(\bigcap_{n=1}^{\infty} \{\omega : \zeta_n \leqslant k-1\}\right)$.

To prove that $\sum P\{A_M\} = 1$, apply the Borel-Cantelli lemma.

353. The sequence ξ_n can be constructed so that for $\sigma_n^2 \leqslant n^2$ the following conditions are satisfied:

$$P\{\xi_n = n\} = P\{\xi_n = -n\} = \frac{\sigma_n^2}{2n^2};$$

$$P\{\xi_n = 0\} = 1 - \frac{\sigma_n^2}{n^2},$$

and for $\sigma_n^2 \leqslant n^2$, the conditions:

$$\tfrac{1}{2} = P\{\xi_n = \sigma_n\} = P\{\xi_n = -\sigma_n\}.$$

Then apply the Borel-Cantelli lemma to the ω-sets $\{\omega : \xi_n(\omega) \geqslant n\}$.

355. It follows from the fact that the sequence $\{1/n \sum_{i=1}^{n} \xi_i\}$ converges to zero with probability 1 that the series $\sum_{n=1}^{\infty} \xi_n/n^{1+\varepsilon}$ converges. The quantities $|\xi_n/n^{1+\varepsilon}| < C < \infty$ and consequently, according to the three series theorem, the series $\sum D[\xi_n]/n^{1+\varepsilon}$ also converges.

365. $M[\xi \mid \mathfrak{A}] = \begin{cases} \dfrac{1}{P(A)} \displaystyle\int_A \xi(\omega) \, P(\mathrm{d}\omega) & \text{for} \quad \omega \in A; \\[4mm] \dfrac{1}{P(\bar{A})} \displaystyle\int_{\bar{A}} \xi(\omega) \, P(\mathrm{d}\omega) & \text{for} \quad \omega \in \bar{A}; \end{cases}$

$M[\chi_B(\omega) \mid \mathfrak{A}] = P\{B \mid \mathfrak{A}\};$

$P\{B \mid \mathfrak{A}\} = \begin{cases} \dfrac{P\{B \cap A\}}{P\{A\}} & \text{for} \quad \omega \in A; \\[4mm] \dfrac{P\{B \cap \bar{A}\}}{P\{\bar{A}\}} & \text{for} \quad \omega \in \bar{A}; \end{cases}$

Answers

367. a) For any $A \in \mathfrak{A}$, since $\xi \geqslant \eta$,

$$\int\limits_A M[\xi \mid \mathfrak{A}] P(d\omega) = \int\limits_A \xi P(d\omega) \geqslant \int\limits_A \eta P(d\omega) = \int\limits_A M[\eta \mid \mathfrak{A}] P(d\omega).$$

Set

$$A_n = \left\{ \omega : M[\xi \mid \mathfrak{A}] < M[\eta \mid \mathfrak{A}] - \frac{1}{n} \right\}.$$

Clearly,

$$A_n \in \mathfrak{A} \text{ and } \int\limits_{A_n} M[\xi \mid \mathfrak{A}] P(d\omega) \leqslant \int\limits_{A_n} M[\eta \mid \mathfrak{A}] P(d\omega) - \frac{1}{n} P\{A_n\}.$$

Consequently, $P\{A_n\} = 0$. But

$$A = \{M[\xi \mid \mathfrak{A}] < M[\eta \mid \mathfrak{A}]\} \quad \bigcup_n A_n.$$

From which it also follows that $P\{A\} = 0$.

c) It follows from the relations $0 \leqslant \xi_1 \leqslant \cdots \leqslant \xi_n \leqslant \cdots$, according to a), that $0 \leqslant M[\xi_1 \mid \mathfrak{A}] \leqslant \cdots \leqslant M[\xi_n \mid \mathfrak{A}] \leqslant \cdots$. For almost all ω the limit of $M[\xi_n \mid \mathfrak{A}]$ exists as $n \to \infty$, possibly infinite; denote it by $\phi(\omega)$. But since the limit of the Lebesgue integral of a monotone sequence of functions equals the integral of the limit function, then, for an arbitrary $A \in \mathfrak{A}$,

$$\lim_{n \to \infty} \int\limits_A M[\xi_n \mid \mathfrak{A}] P(d\omega) = \int\limits_A \phi(\omega) d\omega.$$

On the other hand, according to the definition of conditional mathematical expectation,

$$\int\limits_A \xi_n(\omega) d\omega = \int\limits_A M[\xi_n \mid \mathfrak{A}] d\omega;$$

passing to the limit, we obtain

$$\int\limits_A \phi(\omega) P(d\omega) = \lim_{n \to \infty} \int\limits_A \xi_n(\omega) P(d\omega) = \int\limits_A \lim_{n \to \infty} \xi_n(\omega) P(d\omega) =$$
$$= \int\limits_A \xi(\omega) P(d\omega).$$

It follows that $\phi(\omega) = M[\xi \mid \mathfrak{A}]$.

140

369. $F(\rho) = \int\limits_0^1 r f(\rho, r)\, dr,$ where

$$f(\rho, r) = \begin{cases} \rho^2 & \text{if } 0 \leqslant \rho \leqslant 1 - r; \\ \rho^2 - S_1 + S_2 & \text{if } 1 - r \leqslant \rho \leqslant 1 + r; \\ 1 & \text{if } 1 + r \leqslant \rho; \end{cases}$$

$$S_1 = \frac{\rho^2}{2}\left(\frac{\alpha}{180} - \frac{\sin\alpha}{\pi}\right), \quad \alpha = 2\left(180 - \arccos\left\{\frac{\rho^2 + r^2 - 1}{2r\rho}\right\}\right);$$

$$S_2 = \frac{1}{2}\left(\frac{\beta}{180} - \frac{\sin\beta}{\pi}\right), \quad \beta = 2\arccos\left\{\frac{1 + r^2 - \rho^2}{2r}\right\}.$$

371. Set

$$x_1 = \rho\cos\phi_1;$$
$$x_2 = \rho\sin\phi_1\cos\phi_2;$$
$$x_3 = \rho\sin\phi_1\sin\phi_2\cos\phi_3; \dots;$$
$$x_{n-1} = \rho\sin\phi_1\dots\sin\phi_{n-2}\cos\phi_{n-1};$$
$$x_n = \rho\sin\phi_1\dots\sin\phi_{n-1}$$

and let

$$z \equiv \rho^{n-1}\int\limits_0^\pi \sin^{n-2}\phi_1\, d\phi_1 \int\limits_0^\pi \sin^{n-3}\phi_2\, d\phi_2 \dots$$

$$\dots \int\limits_0^\pi \sin\phi_{n-2}\, d\phi_{n-2} \int\limits_0^{2\pi} h\, d\phi_{n-1};$$

then the conditional density is hz^{-1}; the conditional mathematical expectation is

$$z^{-1}\rho^{n-1}\int\limits_0^\pi \sin^{n-2}\phi_1\, d\phi_1 \dots \int\limits_0^\pi \sin\phi_{n-2}\, d\phi_{n-2} \int\limits_0^{2\pi} hf\, d\phi_{n-1}.$$

7

Infinitely divisible distributions. Normal law. Multidimensional distributions

373. $f(t) = e^{ia}$, where a is an arbitrary constant. Will be.

393. No.

403. The density $p(x, y)$ must depend only on $x^2 + y^2$.

405. This will be the vector with coordinates $c_{i,j} = \sum_{k=1}^{m} p_{i,k} b_k$.

409. $-\frac{1}{2} \leqslant c \leqslant 1$.

411. $\rho(\xi, \eta) = \dfrac{c^2 - a^2 - b^2}{2ab}$.

8   Markov chains

415. b) All the states are essential.

c) One can go from the 2nd state into the 3rd in two steps.

$$P_2 = P^2 = \begin{bmatrix} \frac{13}{36} & \frac{13}{36} & \frac{1}{9} & \frac{1}{6} \\ \frac{5}{12} & \frac{5}{12} & \frac{1}{6} & 0 \\ \frac{5}{24} & \frac{11}{24} & \frac{1}{12} & \frac{1}{4} \\ \frac{1}{4} & \frac{1}{2} & 0 & \frac{1}{4} \end{bmatrix}$$

419. $P_1 = \|p_{ij}\|$,

where

$$p_{ij} = \begin{cases} 0 & j < i - 1, \\ p & j = i - 1, \\ q & j = i, \\ r & j = i + 1, \\ 0 & j > i + 1, \end{cases}$$

$$P_2 = P^2 = \|p_{ij}^2\|,$$

where

$$p_{ij}^{(2)} = \begin{cases} 0 & j < i - 2, \\ p^2 & j = i - 2, \\ 2pq & j = i - 1, \\ q^2 + 2pr & j = i, \\ 2rq & j = i + 1, \\ r^2 & j = i + 2, \\ 0 & j > i + 2. \end{cases}$$

421. No.

423. No.

425. a) $\sum\limits_{k=1}^{\infty} C_i C_k \exp\{-\alpha|i-k| - \alpha|k-j|\}$;

b) $C_i = \dfrac{1 - e^{-\alpha}}{1 + e^{-\alpha} - e^{-\alpha i}}$.

431. $P_1(n) = \alpha p_{11}^{(n)} + \beta p_{21}^{(n)} = \dfrac{1}{2} + \dfrac{(\alpha - \beta)(p - q)^n}{2}$;

$P_2(n) = \alpha p_{12}^{(n)} + \beta p_{22}^{(n)} = \dfrac{1}{2} - \dfrac{(\alpha - \beta)(p - q)^n}{2}$;

$p_1 = p_2 = \frac{1}{2}$.

4.35 $a_0 = a_2 = \frac{1}{3}$; $a_1 = a_3 = \frac{1}{6}$.

Suggested reading

[1] CHUNG, K. L. *Markov chains with stationary transition probabilities.* Springer-Verlag, New York, (1967).

[2] FELLER, W. *An introduction to probability theory and its applications I.* John Wiley, New York, (1957).

[3] GANGOLLI, R. A. and D. YLVISAKER. *Discrete probability.* Harcourt-Brace, New York, (1967).

[4] HOEL, P. G. *Introduction to mathematical statistics.* John Wiley, New York, (1966).

[5] KARLIN, S. *A first course in stochastic processes.* Academic Press, New York, (1966).

[6] KEMENY, J. G. and J. L. SNELL. *Finite Markov chains.* Van Nostrand, New York, (1960).

[7] KEMENY, J. G., J. L. SNELL and A. W. KNAPP. *Denumerable Markov chains.* Van Nostrand, New York, (1966).

[8] KHINCHIN, A. I. *Mathematical foundations of information theory.* Dover, New York, (1957).

[9] KINGMAN, J. F. C. and S. J. TAYLOR. *Introduction to measure and probability.* Cambridge University Press, (1966).

[10] LOÈVE, M. *Probability theory.* Van Nostrand, New York, (1955).

[11] PARZEN, E. *Modern probability theory and its applications.* John Wiley, New York, (1960).

[12] PROHOROV, YU. V. and YU. A. ROZANOV. *Probability theory.* Springer-Verlag, New York, (1969).

[13] TUCKER, H. *A graduate course in probability.* Academic Press, New York, (1967).

[14] VAN DER WAERDEN, B. L. *Mathematical statistics.* Springer-Verlag, New York, (1969).

Index

Index

148